AF577017

Laser Heating of Metals

The Adam Hilger Series on Optics and Optoelectronics

Series Editors: **E R Pike** FRS and **W T Welford** FRS

Other books in the series

Aberrations of Optical Systems
W T Welford

Laser Damage in Optical Materials
R M Wood

Waves in Focal Regions
J J Stamnes

Laser Analytical Spectrochemistry
edited by V S Letokhov

Laser Picosecond Spectroscopy and Photochemistry of Biomolecules
edited by V S Letokhov

Cutting and Polishing Optical and Electronic Materials
G W Fynn and W J A Powell

Prism and Lens Making
F Twyman

The Optical Constants of Bulk Materials and Films
L Ward

Infrared Optical Fibers
T Katsuyama and H Matsumura

Solar Cells and Optics for Photovoltaic Concentration
A Luque

The Fabry–Perot Interferometer
J M Vaughan

Interferometry of Fibrous Materials
N Barakat and A A Hamza

Physics and Chemistry of Crystalline Lithium Niobate
A M Prokhorov and Yu S Kuz'minov

The Adam Hilger Series on Optics and Optoelectronics

Laser Heating of Metals

A M Prokhorov, V I Konov

General Physics Institute, USSR Academy of Sciences, Moscow

I Ursu and I N Mihăilescu

Institute of Atomic Physics, Bucharest

Adam Hilger
Bristol, Philadelphia and New York

British Library Cataloguing in Publication Data

Prokhorov, A. M.
Laser heating of metals.
1. Metals. Processing. Use of lasers
I. Title
671

ISBN 0-7503-0040-X

Library of Congress Cataloging-in-Publication Data are available

Published under the Adam Hilger imprint by IOP Publishing Ltd
Techno House, Redcliffe Way, Bristol BS1 6NX, England
335 East 45th Street, New York, NY 10017-3483, USA

US Editorial Office: 1411 Walnut Street, Philadelphia, PA 19102

Typeset by P&R Typesetters Ltd, Salisbury, England
Printed in Great Britain by J W Arrowsmith Ltd, Bristol

Contents

Series Editors' Preface

Optics has been a major field of pure and applied physics since the mid 1960s. Lasers have transformed the work of, for example, spectroscopists, metrologists, communication engineers and instrument designers in addition to leading to many detailed developments in the quantum theory of light. Computers have revolutionised the subject of optical design and at the same time new requirements such as laser scanners, very large telescopes and diffractive optical systems have stimulated developments in aberration theory. The increasing use of what were previously not very familiar regions of the spectrum, e.g. the thermal infrared band, has led to the development of new optical materials as well as new optical designs. New detectors have led to better methods of extracting the information from the available signals. These are only some of the reasons for having an *Adam Hilger Series on Optics and Optoelectronics.*

The name Adam Hilger, in fact, is that of one of the most famous precision optical instrument companies in the UK; the company existed as a separate entity until the mid 1940s. As an optical instrument firm Adam Hilger had always published books on optics, perhaps the most notable being Frank Twyman's *Prism and Lens Making.*

Since the purchase of the book publishing company by The Institute of Physics in 1976 their list has been expanded into all areas of physics and related subjects. Books on optics and quantum optics have continued to comprise a significant part of Adam Hilger's output, however, and the present series has some twenty titles in print or to be published shortly. These constitute an essential library for all who work in the optical field.

Series Editors' Preface

Preface

This book dwells upon a field of outstanding interest in current science and technology, now in full and dynamic development. The rapid advances with high-intensity laser systems have resulted in a rapid translation of this very important product from research laboratory to production line.

The advent of high-power laser systems—an important event over a quarter of a century ago—was accompanied by the rapid assertion of laser metal processing as their chief application. Nowadays even in traditional realms of technology, such as welding, cutting and drilling, modern lasers with high-quality beams, rated at average powers of the order of 10 kW and more, are used to attain productivity levels comparable with classic methods, while also clearly surpassing these by a unique precision and a highly localised thermal action. Moreover, recently a series of achievements termed laser micrometallurgy—laser-light-induced structural modifications, and chemical compound formation in the surface layers of processed metal samples—have been published.

Experience has shown that large-scale implementation and efficient use of lasers in any field are inconceivable without a thorough knowledge of the fundamental laws governing the interaction of radiation with matter. This conclusion stands true for laser processing of metals as well. Indeed, the approach taken in the 1960s and still practiced in the early 1970s, to rely only on empirical working schemes and prescriptions in the laser processing of various materials, metals included, has ultimately proved a 'dead-end' solution. Among the drawbacks is the fact that the empirical approach, based on trial-and-error attempts, is time and energy consuming, and also shows a limited repeatability of results. The high costs of risk, and the impossibility of promoting optimal design and advanced planning of technological developments in the field were also disadvantages of this approach.

The alternative approach—the fundamentals of which are given in this book—is a systematic and in-depth study of the physical and chemical mechanisms governing the interaction of laser radiation with solid targets, among them metals in different gaseous environments, and for a wide range of beam parameters. Modern methods of investigation (x-ray diffraction, electron microscopy, mass spectrometry, local elemental and structural analysis, ultrafast photography, etc) supported by computer-aided data processing have given considerable impetus to research in the field and have challenged many scientific teams to focus upon providing explanatory, normative and predictive models in order to lay sound, reliable foundations to the new enterprise, while also ensuring its best targeting on relevant objectives. As a result, a great many new phenomena have been revealed, and their adequate interpretation given.

In this book we shall discuss the heating of metals by laser beams at intensity levels too low, or irradiation times too short, to cause target material ablation and plasma formation near the metal surface. Despite this self-imposed limitation, the topics dealt with here enjoy unanimous and well deserved attention, because target heating by radiation is a basic phase in practically all schemes of interaction of laser radiation with matter and in processing technologies.

The specificity of the proposed approach to the problem rests with the parameters governing the heating kinetics under the above mentioned restrictions. Among these are the thermophysical properties of metals; the optical properties of the metal target surface; the beam and target sizes; the distribution in space of the laser beam and its temporal–spatial, intensity, polarisation and angle of incidence; the composition of the surrounding gas, and its pressure.

Simple and convenient formulae are given for calculating the evolution in space and time of the temperature distribution in metal targets for the most common irradiation conditions and target geometries. The temperature variation of the optical and thermophysical properties of the target material, including its transition through the melting point, are given particular emphasis.

Special attention is paid to radiation-induced reversible and irreversible thermodeformations and their contribution, as well as that of metal melting and vaporisation, in the optical damage of laser mirrors.

Two classes of laser-induced surface phenomena are also discussed, which can play an important part in determining the heating rate of the metal target, namely the interaction of the laser beam with self-induced, or pre-existing, resonant and non-resonant surface structures; and processes of thermochemical origin. Although both have received much attention in numerous original publications, a systematic, integrative presentation was thought appropriate, and timely.

Speaking of processes of thermochemical origin, when a target is irradiated in a chemically active gas, e.g. in air or oxygen, the reaction products can be generated in quantities that would suffice to affect, or even to completely determine, the interaction process. Two such kinds of influence will be considered: the variation of the target's optical properties; and the additional heat release in the irradiated metal sample, due to exothermal reactions.

In the light of the authors' claim to give a comprehensive presentation of the basic physical and chemical processes induced by laser irradiation on solid surfaces, their concentration on mainly one theoretical and experimental pattern, namely the interaction of CO_2 laser radiation (wavelength $\lambda \simeq 10\ \mu m$) with metals in oxidising, or chemically neutral, atmospheres, may appear paradoxical at a first glance. Under closer examination it will be seen that such a choice is first of all based on the fact that CO_2 lasers are, at present, the most advanced and the most widely used in metalworking, which can be performed both in air and an inert-gas atmosphere. It is also a fact that, over the limited extent of a single book, it is not possible to present all the relevant information required in order to describe the specific features of the innumerable combinations: radiation wavelength–target material–surrounding gas.

As to the generality of the proposed approach, one can say that the majority of experimental and theoretical results here reported can be employed either directly, or with minimal adaptation, to a variety of irradiation regimes, differing in the value of λ, in target material and in the gaseous environment. Note that some results and models are quite valid not only for semiconductors or insulators, but also for biological tissues. Moreover, we give many references regarding experimental results obtained with other radiation wavelengths, in particular with neodymium lasers ($\lambda = 1.06\ \mu m$).

The book addresses a wide range of topics, thus having the potential to interest researchers specialising in the field of interaction of radiation with condensed matter, university teaching staff, students and experts from industry. This text in no way claims to giving infallible recipes or prescriptions, on how one is to laser-process metals. On the other hand, it is believed that a careful analysis and creative approach to the material presented can help engineers in the field to better understand their problems, and to choose optimal irradiation conditions.

The book is to a large extent based on the authors' results, many of which have been obtained through the fruitful collaboration between the laboratories of the Institute of General Physics of the Academy of Sciences of the USSR, and the National Centre for Physics in Romania. We express true gratitude to our Soviet and Romanian colleagues, for their valuable contribution to the joint scientific research programme on the interaction of laser radiation with solid surfaces, as well as in the preparation of this book.

A M Prokhorov
V I Konov
I Ursu
I N Mihăilescu

List of Symbols and Abbreviations

A	radiation absorptivity (defined as the ratio between the intensity/power/energy absorbed by the sample and the incident intensity/power/energy)
A_{A}	absorptivity due to anomalous skin effect
A_{D}	Drude (model) absorptivity
A'_{D}	effective Drude (model) absorptivity
A_{d}	absorptivity of abrasive particles
A_{e}	effective (integral) absorptivity also called energy coupling coefficient
$A_{\mathrm{e}}^{\mathrm{max}}$	maximum value of the effective absorptivity
A_{ext}	external absorptivity (due to sample surface status)
A_0^{f}	cold absorptivity value after laser heating in air up to a T_{f} final temperature
A_{i}	intrinsic absorptivity
A_i	coefficients of a polynomial describing temperature dependence of absorptivity
A_i^{exp}	experimentally determined values of the above coefficients
A_{IB}	absorptivity due to interband transitions
A_{id}	absorptivity by surface defects and impurities
A_{IR}	absorptivity value obtained by the best fitting of the experimental data in the infrared range
$A_{\mathrm{max}}^{\mathrm{K}}$, $A_{\mathrm{min}}^{\mathrm{K}}$	absorptivity maxima/minima appearing by laser radiation interference inside the oxide–metal system
A^{l}	absorptivity of liquid metal
A_{l}	local (value of) absorptivity
$A_{\mathrm{l}}^{\pm}$	values of the local absorptivity at the interferential maxima/minima between the incident laser wave and the surface electromagnetic waves
$\bar{A}_{\mathrm{l}}$	value of local absorptivity averaged over a period of the surface periodic structure

A_M pure metal absorptivity
A_{md} average value of absorptivity
A_{min} minimum experimentally found absorptivity value of a certain metal at a given wavelength
A_{min}, A_{max} minimum/maximum values of absorptivity during laser heating
A_N absorptivity value according to the normal skin effect
A_{ox} absorptivity of the oxide layer and of the layers consisting of adsorbed substances
A_R actual value of absorptivity due to laser heating and surface condition
A_r absorptivity due to surface roughness
A^s absorptivity of solid metal
A_s absorptivity at the irradiation spot
A_{SEW} supplementary absorptivity due to surface electromagnetic waves induction and propagation on RPS rippled metal surface
A_v fraction of radiation energy absorbed into the sample per unit time and unit volume of metal
A_0 initial absorptivity
A_1, A_2 supplementary absorptivity due to induction and propagation of surface electromagnetic waves inside and outside the irradiation spot, respectively
$A_\perp$ absorptivity of radiation linearly polarised normal to the incidence plane
$A_\parallel$ absorptivity of radiation linearly polarised parallel to the incidence plane
a amplitude of the magnetic field of the surface electromagnetic waves; radius of a dielectric sphere (impurity)
a_m reaction (oxidation) rate
C_V heat capacity
C_V^0 initial value of heat capacity
c specific heat
c_0 speed of light
D_{O_2} diffusion coefficient for oxygen molecules
D_s diameter of the irradiation spot
d (gas) diffusion constant
d_g grain dimension
E_b electron binding energy
$\boldsymbol{E}_i$ electric field of the incident laser wave
E_m threshold fluence, for melting in a surface layer; sometimes bears superscripts specifying the nature of ambient gas where the irradiation is performed
E_s radiation fluence, defined as ratio of the total incident energy, E_0 to the area S_s of the irradiation spot
E_s^m threshold fluence, for melting
E_s^p threshold fluence, for damage

E_s^0	maximum value of radiation fluence
E_v	energy expenditure for the processing of a unit length of material
E_0	total energy of radiation pulse
e	electron (electrical) charge
F_G	figure of merit for damage by melting
F_0	Fourier number
f	amplitude of the electric field of the incident laser wave; repetition rate of the consecutive laser pulses; parameter depending on the surface status
f_T, f_2, f_3	functions of Fourier number
G	displacement modulus
$GJLC$	gas jet laser cutting
h	thickness of (metal) sample; amplitude depth of the surface periodic structures
h_0	optimum value of the amplitude depth of the surface periodic structure as against maximum energy coupling of laser energy to sample
h^*	amplitude depth of the surface periodic structure for which the absorptivity is doubled as against the value corresponding to the bare (plane) surface
I	radiation intensity
I_m	intensity threshold for melting
I_{max}	maximum allowed intensity for which surface distortion is smaller than a certain value specified by the subscript: e.g. δ_{max}, $\lambda/20$, σ_m
I_p	intensity threshold for damage; usually the symbol bears various superscripts specifying the cause of damage
I_p^{pp}	intensity threshold of the periodically pulsed radiation causing damage by cumulative stress
I_{SEW}	intensity of surface electromagnetic waves
I_v	intensity threshold for vaporisation
J_0, J_1	Bessel functions of the first kind and of zero and first orders, respectively
J_0^m	modified Bessel function of zero order
K_0^m	modified Bessel function of the second kind and of zero order
k	absorption index
k_{ox}	thermal conductivity of oxide thermal conductivity
$\bar{k}_T$	average value of the thermal conductivity
k_T	thermal conductivity
k_0	initial value of thermal conductivity
L	transverse dimension of sample
L_m	latent heat of melting
l_c	length of a crack

l_e	mean free path of electron inside metal
l_{SEW}	free path on surface of the surface electromagnetic waves
l_{th}	thermal diffusion depth; diffusion length
M	molecular mass
m	electron mass; electron rest mass
m_e	electron effective mass
Δm, m	mass of (metal) sample
N	number of consecutive laser pulses applied at the same location on the sample surface; volume concentration of ions; oxygen concentration
N_0	oxygen concentration on surface
n	refractive index also called refraction coefficient
n_a	concentration of metal atoms (in solid phase)
n_a^l	concentration of metal atoms in liquid phase
n_c	complex refractive index
n_e	electron density, also called electron concentration
n_e^*	effective electron density
n_0	density of valence electrons; the same symbol is used for the number of conduction electrons per metal atom
n_0^l	the number of conduction electrons per metal atom in liquid phase
P	power of radiation; summary (total) power in the laser oxidation process
P_{ex}	power released by an exothermal (oxidation) reaction
P_l	specific energy (power) consumption per unit length of laser cutting
p	pressure of ambient gas
p_{cr}	critical value as against oxidation of pressure of ambient air
p_{O_2}	pressure at the metal–oxide boundary
Q	power of thermal losses
Q_T	power of thermal losses through thermocouples and connection wires
$Q_\pm$	power of thermal losses during sample heating by laser irradiation and subsequent cooling
q	(specific) latent heat for vaporisation
R	radiation reflectivity (defined as the ratio of the intensity/power/energy reflected by the sample to the incident intensity/power/energy)
R_D	diffusion (scattering) reflectivity; the same symbol is used for Drude (model) reflectivity
R_i	intrinsic reflectivity
R_N	reflectivity value according to the normal skin effect
R_R	specular reflectivity
R_R^0	initial value of specular reflectivity

R_r	reflectivity of a rough surface
R_s	radius of the irradiation spot
R_s^e, $R_s^{e^2}$	radii within the irradiation spot at which the radiation intensity/fluence is e and e^2 times lower than at the beam axis (the maximum value)
$R_\perp$	reflectivity of radiation linearly polarised normal to the incidence plane
$R_\parallel$	reflectivity of radiation linearly polarised parallel to the incidence plane
r	current radius; incidence angle corresponding to resonant excitation of surface electromagnetic waves
r_f	final value of electrical resistivity
r_{ij}	amplitude coefficients, of radiation reflection at various interfaces
r_M	electrical resistivity of (pure) metal
r_0	(electrical) resistivity of pure metal; initial value of electrical resistivity
S	total area of sample surface
S_s	area of irradiation spot
SB	Stefan–Boltzmann constant
SEW	surface electromagnetic waves
SPS	surface periodic structures
T	current temperature
$\bar{T}$	average temperature
ΔT	temperature excursion
T_a	activation temperature of metal oxidation
T_b	burning temperature (of metal)
T_d	activation temperature of (gas) diffusion; temperature of diffusion activation
T_{dW}	Wagner activation temperature of diffusion
T_f	final (value of) temperature
T_G	temperature reached on the surface under the action of radiation exhibiting a Gaussian intensity/energy spatial distribution
T_g	temperature reached on the surface under the action of a radiation pulse having a Gaussian time-shape
T_{ig}	ignition temperature of metal
T_m	melting point (temperature)
T_s	surface temperature
T_{st}	stationary value of temperature
T_v^{ox}	initiating temperature of intense vaporisation of the oxide
T_0	initial value of temperature
T_1, T_2	temperature values at the air/oxide interface and at the metal/oxide interface, respectively

T^+, T^-	temperature values at the interference maxima/minima between the incidence laser wave and the surface electromagnetic waves
t	current time
t_a	duration of the activation stage of oxidation
t_f	time moment at the end of irradiation
t_m	time of melting
t_{st}	time elapsed since a stationary value of temperature is reached on surface
t_z	time after which the melting is reached of a layer of thickness z
u	dimensionless variable
u_0	sound velocity inside solid
v	scanning velocity of laser beam across the sample surface
v_a	ablation velocity of sample under laser irradiation
v_e	mean electron velocity inside metal
v_F	velocity of the conduction electrons
v_{opt}	optimum value of scanning velocity as against laser burning of sample with minimum energy/power consumption
v_{ox}	oxidation rate
W_H	latent heat of the (oxidation) reaction
x	thickness of oxide layer; current coordinate
x_f	final value of the oxide layer thickness
x_k	thickness of oxide layer characterising the influence (strong/weak) of the electric field upon the oxidation process
x^k_{max}, x^k_{min}	oxide thickness values for which absorptivity maxima/minima are observed in the oxide–metal system
x_T	target dimension
x_0	initial value of the oxide layer thickness
Y_0, Y_1	Bessel functions of the second kind and of zero and first orders, respectively
Z	(complex) impedance of the optical surface; the same symbol is used for the impedance of the oxide layer
Z_1	real part of the impedance of the optical surface
Z_2	imaginary part of above
z	surface profile; coordinate directed inward the sample
z_m	position of solid/molten metal interface, i.e. the thickness of the (surface) molten layer
α	absorption coefficient of radiation; attenuation coefficient of surface electromagnetic waves
α_d	attenuation coefficient of surface electromagnetic waves by heat dissipation into the (metal) substrate
α^e	attenuation coefficient of surface electromagnetic waves outside the irradiation spot (i.e. on the 'cold' planar surface)
α_p	electric polarisability

α_p^0 the direct current electrical polarisability
α_r attenuation coefficient of surface electromagnetic waves by radiation in ambient vacuum gas
α_T coefficient of linear dilation
β linking coefficient between the incident laser wave and the surface electromagnetic waves
Γ electron collision frequency also called electron relaxation frequency
Γ_{ed} electron–defect collision frequency
Γ_{ee} electron–electron collision frequency
Γ_{eff} effective (electron) collision frequency
Γ_{ep} electron–phonon collision frequency
Γ_{ep}^{cl} classical (as determined by direct current) value of above
Γ_p concentration of surface impurities/defects
γ_E Euler constant
δ characteristic roughness of surface; distortion of surface submitted to irradiation, also called thermodeformation amplitude; skin layer depth
δ_A skin layer depth according to the anomalous skin effect
δ_D Drude (model) depth of the skin layer
δ_l roughness space correlation over large distances
δ_{max} maximum value of thermodeformation amplitude
δ_N skin layer depth according to the normal skin effect
δ_s roughness space correlation over short distances
δ_{SEW} attenuation depth of the surface electromagnetic waves
δ_0 skin layer depth (of metal)
ε dielectric permittivity also called dielectric constant (at the radiation wavelength)
ε_c complex dielectric permittivity
ε_1 real part of the complex dielectric permittivity
ε_M dielectric permittivity of the (base) metal
ε_2 imaginary part of the complex dielectric permittivity
ε_∞ high-frequency dielectric permittivity
η constant of convective heat exchange also called convective losses constant
θ incidence angle
Λ period of the surface periodic structures
λ radiation wavelength
λ_{IB} limiting wavelength below which the interband absorption becomes significant
λ_{min} radiation wavelength for which a minimum absorptivity (value) is recorded
λ_p wavelength of a probing laser beam
μ carrier mobility

ν	Poisson's coefficient
ξ	roughness profile (on surface)
ρ	density (metal) or oxide
ρ_{O_2}	oxygen density
σ	(electrical) conductivity
σ_B	strength modulus
σ_{ed}	cross section of electron scattering on impurities
σ_f	friction stress
σ_{ik}	stress tensor components
σ_m	thermoplastic stress causing irreversible damage also called the elasticity threshold of metal
σ_p	energy of plastic deformation of the surface; average roughness over long distances
σ_T	flow limit of metal, i.e. the threshold for plastic deformation (generally) causing slip-banding
σ_y	fatigue stress
σ_0	the direct current electrical conductivity; surface emissivity, also called blackening degree
σ_0^+	surface emissivity during (after) laser heating and subsequent cooling
σ_1	uniaxial stress
τ	period of electron collisions, also called collision time or average relaxation time
τ_D	delay time
τ_e	etching time
τ_{ef}	effective collision time
τ_p	laser pulse duration
$\Delta\varphi$	potential of the thermoelastic stresses
χ	thermal diffusivity
ω	circular frequency of radiation

Chapter 1 Optical Properties of Metals

The laser heating of a material is mainly determined by that material's absorptivity at the respective laser wavelength. This is why the metal absorptivity—or, alternatively, the reflectivity of the metal at given wavelengths—stands as the chief criterion guiding any approach to an efficient processing of metallic parts. This chapter links the key quantities of absorptivity/reflectivity to the optical properties of metals, the general condition of their surfaces and the temperature, i.e. the laser heating rates.

The heating rate of a metal sample is mainly determined by the sample absorptivity for a given wavelength—a quantity which, in turn, is determined by the optical properties of the metal itself and of the sample surface, as well as by the temperature range, heating rate, etc. That is why the metal absorptivity, A, or alternatively the metal reflectivity, R, stand as chief criteria guiding the choice of the most appropriate laser system for processing metallic parts—by surface hardening, welding, size adjustments, etc [1].

Quantitatively, the absorptivity A is the ratio of the intensity absorbed by the sample, I_a, to the incident intensity, I, at a certain moment in the process of laser heating. Accordingly, the reflectivity, $R = 1 - A$, is the ratio between the reflected (specularly and/or diffused) intensity, I_r, and the incident intensity, I.

The absorptivity of metals shows a general trend to increase when the incident radiation wavelength decreases from the infrared to the ultraviolet spectral range [2, 3]. Generally, for radiation with a wavelength $\lambda \simeq 10.6\ \mu m$ the absorptivity of metals is very low—of the order of a few per cent, and sometimes even of a few fractions of one per cent. As a consequence, in practice metals are largely used as the preferred raw material for manufacturing mirrors to be used in conjunction with CO_2 lasers and other laser systems generating in the mid- and near-infrared spectral ranges. At wavelengths below 1 μm, metals of sufficiently high reflectivity are no longer available—perhaps with the exception of aluminium—and consequently mirrors are usually manufactured by coating a base material with appropriate dielectric films.

At a first glance the low absorptivity values make the use of power CO_2 lasers in metal processing very inefficient. However, it has been demonstrated that the absorptivity can reach significantly higher levels—in some cases it can increase by more than one order of magnitude—during the laser action, due to the increase in temperature as well as to several other processes.

Accordingly, the investigation of the dynamics of variation of absorptivity during the process of laser irradiation is one topic of paramount scientific and practical interest.

Two main methods may be used for the experimental determination of absorptivity, namely

(i) direct measurement of A by calorimetric methods; and
(ii) the indirect determination of A based upon measuring the reflectivity.

In the next chapter various methods for the determination of A and R will be described in detail. For the time being, let us mention only that the second method generally encounters several difficulties. First, we note that the signal reflected by a surface actually consists of a specular component and also of a scattered (diffuse) component—so that the value of the total reflection coefficient is $R = R_R + R_D$, where R_R and R_D are the coefficients of specular reflection and scattering, respectively. In order to obtain correct estimations of the R values one has to collect the signal over the whole irradiated surface. Secondly, in the case of the CO_2 laser, pure metals generally exhibit a reflectivity close to unity. As a consequence, minute absolute errors in the determination of R would result in large errors in evaluating A. For example, an almost indiscernible difference between $R = 0.98$ and $R = 0.99$ would result in a difference by a factor of two in the value of the absorptivity—from $A = 0.02$ down to $A = 0.01$.

Nevertheless, the sample reflectivity $R_R + R_D$ and, in some cases, of $R = R_R$ alone, gives valuable information on processes evolving on metallic surfaces during laser irradiation [4–9]. As an illustration, figure 1.1 [4] displays CRO traces of the variation in R_R during the action of the radiation of neodymium laser pulses ($\lambda = 1.06$ μm) upon polished plates of Cu and Ta (the first experiments of this kind were reported in references [5, 6]). From this figure one can appreciate how important the changes in the optical properties of the metals during the process of laser irradiation could be (cf also references [7–9]).

To get a better insight into this issue of critical importance, let us now discuss, at some length, the main theoretical models and experimental data on the reflectivity (absorptivity) of metals, over a wide range of temperatures.

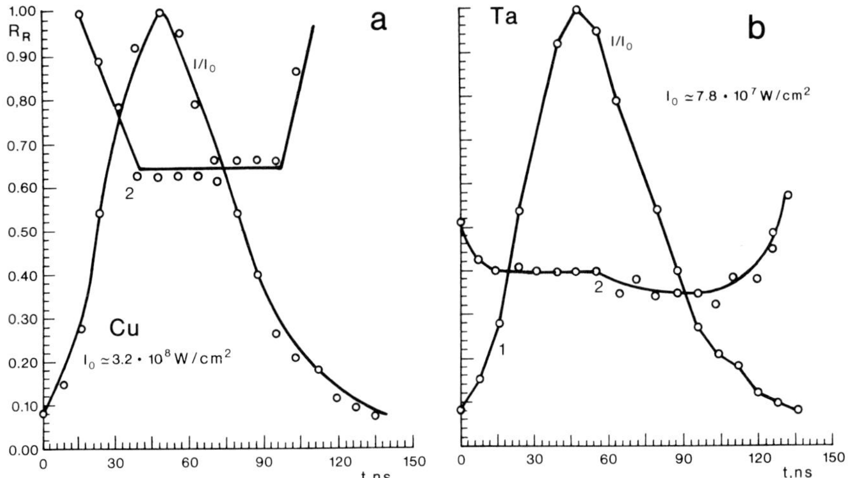

Figure 1.1 Shape of the laser pulse I/I_0 ($\lambda \simeq 1.06\ \mu$m) and the modification during the laser irradiation process of the reflectivity of prepared samples of copper (a) and tantalum (b), respectively. Curves 1 show the temporal profile of the laser pulse and curves 2 the temporal variation of reflectivity.

1.1. Metal absorptivity at room temperature

1.1.1. The Drude formalism

One model, proposed by Drude [10], is based on the assumption that in metals the energy absorption proceeds through the intermediary of the electrons, which further transmit, through collisions, their energy to the crystalline lattice. Collisions would feature a period

$$\tau = 1/\Gamma \tag{1.1}$$

where $\Gamma = \Gamma_{ep} + \Gamma_{ee} + \Gamma_{ed}$, where Γ_{ep}, Γ_{ee} and Γ_{ed} correspond to electron collisions with phonons, other electrons and impurities (defects), respectively. The equation describing the electron motion would then read

$$m^* \frac{d^2 y}{dt^2} + m^* \Gamma \frac{dy}{dt} = -e|\boldsymbol{E}_i| \exp(-i\omega t). \tag{1.2}$$

Here y is the coordinate that gives the electron position. The electric vector of the incident light wave, of amplitude $\boldsymbol{E}_i$, is directed along the $0y$ axis; ω is the circular frequency of the radiation ($\simeq 1.78 \times 10^{14}$ rad s^{-1} in case of CO_2 laser radiation), m^* is the effective mass of the electron in the metal and e denotes the electron charge. The effective mass m^* can be evaluated with the expression

$$m^* = m\,\frac{n_0}{n_e}, \tag{1.3}$$

where n_0 is the density of valence electrons, n_e the density of free electrons in the metal and $m \simeq 9.11 \times 10^{-31}$ kg is the electron rest mass.

We should mention that other authors use a different expression for the effective mass

$$n_e^* = n_e\,\frac{m}{m^*} \tag{1.4}$$

where the n_e^* parameter, also known as the 'effective' electron density, has, however, no straightforward physical meaning. We shall not use this expression.

The solution of equation (1.2) is

$$y = \frac{e}{m^*}\,\frac{|\boldsymbol{E}_i|\exp(i\omega t)}{\omega^2 - i\omega\Gamma}. \tag{1.5}$$

The electric conductivity, σ, is defined as the ratio of the component of the electric current density oscillating in phase with the electric field, to the intensity of the electric field, i.e. as the real part of the quantity

$$-\frac{n_e e}{|\boldsymbol{E}_i|\exp(i\omega t)}\,\frac{dy}{dt}. \tag{1.6}$$

Thus by using equations (1.5) and (1.6) one obtains

$$\sigma = \frac{n_e e^2 \Gamma}{m^*(\omega^2 + \Gamma^2)}. \tag{1.7}$$

Similarly, the polarisability, α_p, is obtained as the ratio between the current component outphased at 90° with respect to the field (or the imaginary part of the ratio (1.6)) and the circular frequency of the radiation, ω.

Consequently

$$\alpha_p = -\frac{n_e e^2}{m^*(\omega^2 + \Gamma^2)}. \tag{1.8}$$

The direct current (DC) values of these quantities ($\omega = 0$) are

$$\sigma_0 = \frac{n_e e^2}{m^* \Gamma} \tag{1.9}$$

$$\alpha_p^0 = -\frac{n_e e^2}{m^* \Gamma^2}. \tag{1.10}$$

Equation (1.9) is also known as the Lorentz–Sommerfeld relation. Let us mention that equations (1.9) and (1.10) would function well not only at $\omega = 0$, but also whenever the condition $\omega^2 \ll \Gamma^2$ is fulfilled.

The complex dielectric permittivity of a metal can be obtained as

$$\varepsilon_c = \varepsilon_\infty + 4\pi\alpha_p + \frac{4\pi\sigma}{i\omega} \tag{1.11}$$

where $\varepsilon_\infty = 1$ is the dielectric constant at high frequencies ($\omega \to \infty$). Substituting for σ and α their respective expressions (1.7) and (1.8) one obtains

$$\varepsilon_c \simeq 1 - \frac{\omega_p^2}{\omega^2 + \Gamma^2} - i\,\frac{\Gamma\omega_p^2}{\omega(\omega^2 + \Gamma^2)} \tag{1.12}$$

with ω_p denoting the plasma frequency given by

$$\omega_p^2 = \frac{4\pi n_e e^2}{m^*}. \tag{1.13}$$

The order of magnitude of ω_p for metals at room temperature is $\omega_p \sim 10^{16}$ rad s^{-1}, a small deviation from this value resulting from the volume dilation as an effect of heating.

The complex dielectric permittivity ε_c can be written as

$$\varepsilon_c = \varepsilon_1 - i\varepsilon_2 \tag{1.14}$$

where

$$\varepsilon_1 = 1 - \frac{\omega_p^2}{\omega^2 + \Gamma^2} \tag{1.15}$$

and

$$\varepsilon_2 = \frac{\Gamma\omega_p^2}{\omega(\omega^2 + \Gamma^2)}. \tag{1.6}$$

This result can also be obtained from the Boltzmann equation, assuming an average relaxation time $\tau = \Gamma^{-1}$ for the distribution function of the electrons. Concerning the values of these parameters, it is to be noted that, at $\lambda = 10.6\,\mu$m, ε_1 is negative while ε_2 is positive, both exhibiting relatively high absolute values at room temperature (see table 1.1). Finally, it is to be mentioned that the halfwidth of the curve $\varepsilon_2(\omega)$ is precisely Γ.

The complex refractive index of the metal n_c and its reflectivity at normal incidence results from the Fresnel equations

$$n_c = \sqrt{\varepsilon_c} = n + ik \tag{1.17}$$

$$R_D = \left|\frac{n_c - 1}{n_c + 1}\right|^2 = \frac{(n-1)^2 + k^2}{(n+1) + k^2} \tag{1.18}$$

Table 1.1 Values of the parameters n, k, $-\varepsilon_1$ and ε_2 for several pure metals.

Metal	n	k	$-\varepsilon_1$	ε_2
Ag	5.88	76.1	5.8×10^5	9.08×10^2
Al	34.30	108.0	10.54×10^3	7.44×10^3
Au	8.57	75.9	5.68×10^3	1.32×10^3
Co	7.70	31.3	9.26×10^2	4.84×10^2
Cu	12.80	64.0	3.96×10^3	1.67×10^3
Fe	7.60	27.0	6.74×10^2	4.12×10^2
Ni	7.44	39.2	1.49×10^3	5.89×10^2
Pb	22.30	38.5	9.82×10^2	1.72×10^3
Pd	6.60	21.9	4.37×10^2	2.89×10^2
Pt	11.13	39.6	1.47×10^3	8.95×10^2
Ti	8.24	19.4	3.01×10^2	3.19×10^2
W	9.00	44.4	1.89×10^3	8.02×10^2

where n and k are the refractive and absorption indexes of the radiation in the metal, respectively (n is also known as refraction coefficient). From (1.12) and (1.17) one obtains

$$n^2 - k^2 = 1 - \frac{\omega_p^2}{\omega^2 + \Gamma^2} = \varepsilon_1 \tag{1.19}$$

and

$$nk = \frac{\Gamma\omega_p^2}{2\omega(\omega^2 + \Gamma^2)} = \tfrac{1}{2}\,\varepsilon_2 \tag{1.20}$$

which, together with equation (1.9), represent the most commonly known expression of Drude's results, relating the optical to the electric properties of the metal. After some algebra, from (1.19) and (1.20) one obtains

$$n = \frac{1}{\sqrt{2}}\left[\left((1-Q)^2 + \frac{Q\Gamma}{\omega}\right)^{1/2} - Q + 1\right]^{1/2} \tag{1.21}$$

$$k = \frac{1}{\sqrt{2}}\left[\left((1-Q)^2 + \frac{Q\Gamma}{\omega}\right)^{1/2} + Q - 1\right]^{1/2} \tag{1.22}$$

where

$$Q = \frac{\omega_p^2}{Q^2 + \Gamma^2}. \tag{1.23}$$

For many metals, e.g. Ag, Au,

$$\omega_p^2 \gg \omega^2 \gg \Gamma^2. \tag{1.24}$$

From equations (1.18), (1.21) and (1.22) one gets

$$R_{\mathrm{D}}=1-\frac{2\Gamma}{\omega_{\mathrm{p}}} \tag{1.25}$$

and

$$A_{\mathrm{D}}=1-R_{\mathrm{D}}=\frac{2\Gamma}{\omega_{\mathrm{p}}}. \tag{1.26}$$

In other cases (Al, Cu, etc)

$$\omega_{\mathrm{p}}^{2} \gg \omega^{2} \sim \Gamma^{2} \tag{1.27}$$

and one must resort to more difficult transformations in order to obtain analytical expressions for the reflectivity R_{D} and the absorptivity A_{D} of the metal. These can be obtained from the equations (1.17), (1.18) and the following approximation of the relation (1.12)

$$\varepsilon \simeq -\frac{\omega_{\mathrm{p}}^{2}}{\omega^{2}+\Gamma^{2}}-\mathrm{i}\,\frac{\Gamma\omega_{\mathrm{p}}^{2}}{\omega(\omega^{2}+\Gamma^{2})}. \tag{1.28}$$

Since such calculations may prove quite difficult, a simpler formula is often used, inferred on starting from the impedance of the optical surface, Z [10–12]:

$$Z=Z_{1}+\mathrm{i}Z_{2}=(1+\mathrm{i})\left(\frac{2\pi\omega}{c_{0}^{2}\sigma_{0}}\right)^{1/2}(1+\mathrm{i}\omega\tau)^{1/2} \tag{1.29}$$

where

$$Z_{1}=\left(\frac{2\pi\omega}{c_{0}^{2}\sigma_{0}}\right)^{1/2}[(1+\omega^{2}\tau^{2})^{1/2}-\omega\tau]^{1/2} \tag{1.30}$$

$$Z_{2}=\left(\frac{2\pi\omega}{c_{0}^{2}\sigma_{0}}\right)^{1/2}[(1+\omega^{2}\tau^{2})^{1/2}+\omega\tau]^{1/2}. \tag{1.31}$$

Then the reflectivity R_{D} can be obtained as

$$R_{\mathrm{D}}=1-\frac{c_{0}}{\pi}Z_{1}=1-\left(\frac{2\omega}{\pi\sigma_{0}}\right)^{1/2}[(1+\omega^{2}\tau^{2})^{1/2}-\omega\tau]^{1/2}. \tag{1.32}$$

A satisfactory approximation of the reflectivity when $n>1$, $k>1$ [13, 14] is

$$R_{\mathrm{D}} \simeq \exp\left(-\frac{4n}{n^{2}+k^{2}}\right) \simeq \exp\left[-\left(\frac{2\omega}{\pi\sigma_{0}}\right)^{1/2}[(1+\omega^{2}\tau^{2})^{1/2}-\omega\tau]^{1/2}\right]. \tag{1.33}$$

Since for metals in the infrared one has $n \gg 1$, $k \gg 1$ (table 1.1), this approximation features, within the given wavelength range, an error $\leqslant 0.001$

as compared with the values of the reflectivity calculated with the exact formula (1.18) in the case when condition (1.27) is fulfilled. It can also be noticed that, if

$$\frac{4\pi\sigma_0}{\omega(1+\omega^2\tau^2)} \geqslant 1 \tag{1.34}$$

then equation (1.33) is equivalent to equation (1.32).

Under the same basic assumptions $\omega_p^2 \gg \Gamma^2, n^2+k^2+2n \gg 1$, the following expression is established [11, 15]:

$$A_D \simeq \frac{2\sqrt{2}(\omega/\omega_p)[(1+\omega^2\tau^2)^{1/2}-1]^{1/2}}{1+\sqrt{2}(\omega/\omega_p)[(1+\omega^2\tau^2)^{1/2}-1]^{1/2}} \tag{1.35}$$

which—bearing in mind that for metals at $\lambda = 10.6\ \mu\text{m}$ the quantity $\sqrt{2}(\omega/\omega_p)[(1+\omega^2\tau^2)^{1/2}-1]^{1/2}$ is very small—takes the simpler form

$$A_D \simeq 2\sqrt{2}\,\frac{\omega}{\omega_p}\,[(1+\omega^2\tau^2)^{1/2}-1]^{1/2} \tag{1.36}$$

or

$$R_D \simeq 1-2\sqrt{2}\,\frac{\omega}{\omega_p}\,[(1+\omega^2\tau^2)^{1/2}-1]^{1/2}. \tag{1.37}$$

It is also to be mentioned that, in the limit $\omega^2 \gg \Gamma^2$, also known as the extreme relaxation limit, the expression (1.32) for the reflectivity is reduced to equation (1.25).

Within the limits of application of the Drude formalism, the absorption depth of the laser radiation into a metal—also known as the skin depth of the metal—can be evaluated using the expression

$$\delta_D = \frac{c_0}{2\omega_p}\,\frac{[2(1+\omega^2\tau^2)]^{1/2}}{[(1+\omega^2\tau^2)+1]^{1/2}} \tag{1.38}$$

which, with $\Gamma^2 \ll \omega^2$, becomes $\delta_D \simeq c_0/2\omega_p$.

In concluding this section it has to be mentioned that the range of applicability of the Drude mode, also called the relaxation region, corresponds to cases when the electron current induced at a certain site and at a certain time depends not only on the electric field applied at that moment and point, but also cumulates the influence of the electric field prior to that moment. Such an electron inertia, resulting in the aforementioned relaxation effects, is incorporated in the Drude model through the first term in the right-hand side of equation (1.2).

We have started with the presentation of this model because, as we shall see later, it provides a good description of the absorption by metals of the radiation of CO_2 laser sources (at $\lambda \simeq 10.6\,\mu$m). To complete the picture a situation is presented in which the value of the current induced in the metal at a certain site and at a certain time is strictly determined by the instantaneous value of the electric field applied, a case which corresponds to the normal skin effect.

1.1.2. The normal skin effect

At low frequencies ($\omega \ll \Gamma$), the equation of motion of the electron takes the form

$$m^*\Gamma \frac{\mathrm{d}y}{\mathrm{d}t} \simeq -e|\boldsymbol{E}_\mathrm{i}| \exp(\mathrm{i}\omega t). \tag{1.39}$$

Accordingly, from equations (1.15) and (1.16) one obtains

$$\varepsilon_1 \simeq 1 - \frac{\omega_\mathrm{p}^2}{\Gamma^2} \tag{1.40}$$

$$\varepsilon_2 \simeq \frac{\omega_\mathrm{p}^2}{\omega\Gamma} \tag{1.41}$$

and from equations (1.21)–(1.23)

$$n \simeq k \simeq \left(\frac{\omega_\mathrm{p}^2}{2\omega\Gamma}\right)^{1/2} \simeq \left(\frac{2\pi\sigma_0}{\omega}\right)^{1/2}. \tag{1.42}$$

When $\omega_\mathrm{p}^2 \gg \Gamma^2, \omega^2$, from equation (1.42) (known as the Hagen–Rubens relation) as well as from the Fresnel formula (1.18) it follows that

$$R_\mathrm{N} \simeq 1 - 2\left(\frac{2\Gamma\omega}{2}\right)^{1/2} = 1 - \left(\frac{2\omega}{\pi\sigma_0}\right)^{1/2} \tag{1.43}$$

and

$$A_\mathrm{N} \simeq \left(\frac{2\omega}{\pi\sigma_0}\right)^{1/2}. \tag{1.44}$$

Correspondingly, the depth of the skin layer is given by

$$\delta_\mathrm{N} = \frac{c_0}{\omega_\mathrm{p}}\left(\frac{2\Gamma}{\omega}\right)^{1/2} = \frac{c_0}{(2\pi\sigma_0\omega)^{1/2}}. \tag{1.45}$$

The range of application of the normal skin effect approach—which is a particular case of the Drude formalism—can easily be inferred from figure 1.2 [16], where the ratios n/λ and k/λ are represented as computed on the basis of the two models, versus the radiation wavelength, λ, for a metal exhibiting a high electrical conductivity (Ag) and a low electrical conductivity (Hg), respectively. From figure 1.2 one can notice that for

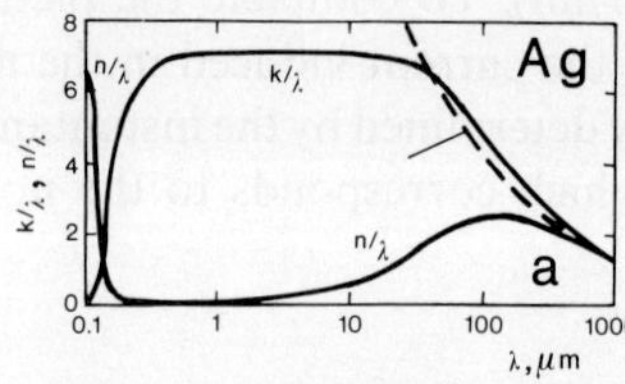

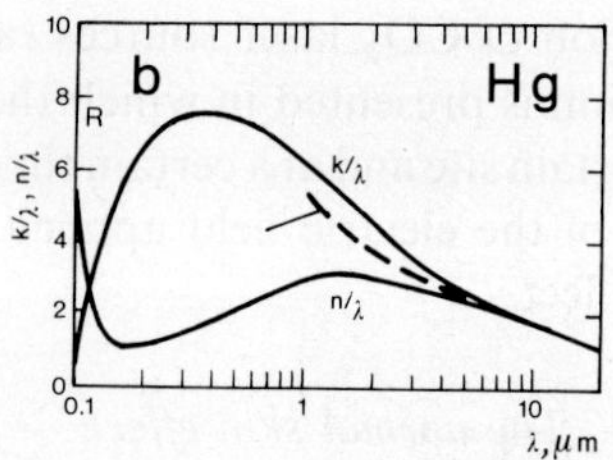

Figure 1.2 The dependence of the ratios n/λ and k/λ on the radiation wavelength, λ, for silver (a) and mercury (b).

Ag—and also for other good electrical conductors—the n and k values become mutually comparable, and also comparable with the values computed on the basis of the normal skin effect, only in the range of very large radiation wavelengths, $\lambda \geqslant 500\,\mu$m. For metals with low electrical conductivity, the application range of the normal skin effect extends to the mid-infrared range (for Hg, this model can be applied for $\lambda \geqslant 7\,\mu$m).

1.1.3. Anomalous skin effect

Since Γ decreases with temperature one can expect (cf equation (1.25)) that, at very low T, the metal absorbtivity would approach the zero limit. However, experiments performed in the 1930s indicated that even the transition to superconductivity was not accompanied by a substantial decrease in A. Subsequently it was established that this $A(T)$ behaviour—called the anomalous skin effect and which cannot be accommodated within the classical approach—occurs whenever the mean free path of the electrons becomes comparable with the radiation wavelength and the penetration depth. Under these circumstances the electron current at a certain point is determined not only by the local value of the electric field, but also by the motion of electrons which could reach that point travelling from distances smaller than, or comparable with, the mean free path. In other words, the speed of a given electron at a certain position depends upon the value of the electric field in other locations situated at distances not exceeding the free electron path.

These effects originate changes in the penetration depth, sample reflectivity and absorptivity that depart from the values predicted according to the Drude theory. Without going through the elaborate calculations required, which can be found in several, well known textbooks [11–13, 16–22], let us introduce the expressions which enable one to evaluate the additional absorptivity, A_A, appearing as a consequence of the anomalous skin effect (1.46) as well the corresponding absorption depth (1.48), in the case of a crystalline lattice featuring a spherical Fermi surface:

$$A_A = \frac{3}{4} \frac{\langle v_F \rangle}{c_0(1+\Gamma^2/\omega^2)}(1-f) + \frac{\omega_p^2}{2\omega^2}\frac{\langle v_F \rangle}{c_0^3} f \tag{1.46}$$

where

$$\langle v_F \rangle = \left(\frac{h}{m^*}\right)\left(\frac{3n_e}{8\pi}\right)^{1/3} \tag{1.47}$$

and

$$\delta_A = \left(\frac{c_0^2 l_e}{2\pi\sigma_0\omega}\right)^{1/3} \simeq \left(\frac{m^* c_0 \langle v_F \rangle}{2\pi n_e e^2 \omega}\right)^{1/3} \simeq \left(\frac{m^* c_0^2 v_e}{2\pi n_e e^2 \omega}\right)^{1/3}. \tag{1.48}$$

Here $\langle v_F \rangle$ and c_0 are the mean electron velocity on the Fermi surface and the light velocity, respectively, $l_e = v_e \tau$ is the mean free path of the electrons in the metal, $v \simeq \langle v_F \rangle \sim 10^8 \text{ cm s}^{-1}$ is the average velocity of the electrons, and $0 \leqslant f \leqslant 1$ is a parameter which depends upon the status of the surface of the metal target.

In reference [23] it was indicated that, for surfaces exhibiting a characteristic roughness δ_σ which does not exceed 30 Å—a requirement on the surface condition which can be met only with the aid of the most advanced methods of surface processing of metal mirrors—it can be taken that $f \simeq 1$, while for an average roughness $\delta_\sigma \geqslant 100$–200 Å, one can safely assume $f \simeq 0$. However, it was shown in [24], based on high-accuracy measurements of reflectivity, that with very high-purity silver films, vacuum coated on highly processed substrates covered with an intermediary layer of calcium fluoride, the transition to $f \simeq 0$ occurs earlier for $\delta_\sigma \simeq 45$–65 Å (figure 1.3).

We may therefore conclude that, for most of the samples which exhibit a roughness often exceeding $\delta_\sigma \simeq 100$ Å, one can take $f \simeq 0$. Thereby, in calculating A_A, the simpler approach described above is to be used

$$A_A \simeq \frac{3}{4}\frac{\langle v_F \rangle}{c_0(1 + \Gamma^2/\omega^2)} \tag{1.49}$$

which as the collision frequency diminishes (i.e. the temperature decreases) and the condition $\Gamma^2 \ll \omega^2$ stands valid, becomes even simpler:

$$A_A \simeq \frac{3}{4}\frac{\langle v_F \rangle}{c_0}. \tag{1.50}$$

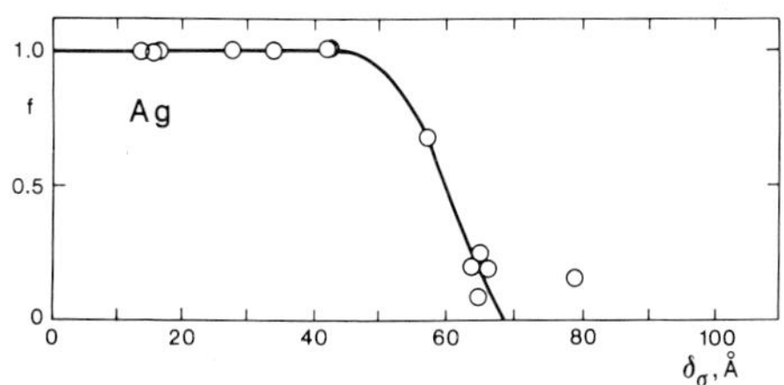

Figure 1.3 Parameter f as a function of the roughness value δ_σ, of a silver surface.

1.1.4. Interband transitions

The behaviour of an electron system has been discussed so far on the assumption that transitions occur only within the same band. Nevertheless, at sufficiently high frequencies, ω, the transitions between the electronic states belonging to different energy bands have an important influence on the optical properties of metals.

The interband transitions can be treated with the Drude formalism by changing equations (1.15) and (1.33) in the following manner:

$$\varepsilon_1 = n^2 - k^2 = 1 - \frac{\omega_p^2}{\omega^2 + \Gamma^2} + \delta\varepsilon \tag{1.51}$$

$$R_D \simeq \exp\left\{\left[-\left(\frac{2\omega}{\pi\sigma_0}\right)^{1/2} [(1+\omega^2\tau^2)^{1/2} - \omega\tau]^{1/2}\right] \times [\![1 + \delta\varepsilon\{[\omega(1+\omega^2\tau^2)^{1/2} + 2\omega^2\tau]/8\pi\sigma_0\}]\!]\right\}. \tag{1.52}$$

The parameter $\delta\varepsilon$ is related to the electric polarisability α_p that is not due to the free carriers, through the equation

$$\delta\varepsilon = 4\pi\alpha_p. \tag{1.53}$$

It is also to be emphasised that the existence of certain interband transitions does not change the expressions (1.16) and (1.20).

Details regarding the values of the quantity $\delta\varepsilon$ in different concrete situations can be found, for example, in reference [25].

1.1.5. The total absorptivity and the regions of validity of the models

The electron path into the skin layer is schematically represented in figure 1.4(*a*) for the following limit situations:

(A) normal skin effect ($\delta \gg l_e$ and $\omega \ll \Gamma$);
(B) relaxation region (Drude formalism applicable) ($\Gamma \leqslant \omega$ and $l_e \ll \delta$);
(D) anomalous skin effect ($l_e \gg \delta$ and $v_e/\omega \gg \delta$†);
(E) extreme anomalous skin effect (also called anomalous reflection) ($v_e/\omega \ll \delta \ll l_e$).

These regions are also represented on a logarithmic diagram ω–τ in figure 1.4(*b*) on which the C transparency zone is also marked.

Indicative diagrams of this type can be drawn for all metals. As an example, the position of copper is indicated on such a diagram in figure 1.5 for two largely differing temperatures. It is noted that at the wavelength of the CO_2 lasers at room temperature we are in the relaxation (B) zone, where the

† The case $v_e/\omega \gg \delta$ and $l_e \geqslant \delta$ is sometimes known as the weak anomalous effect.

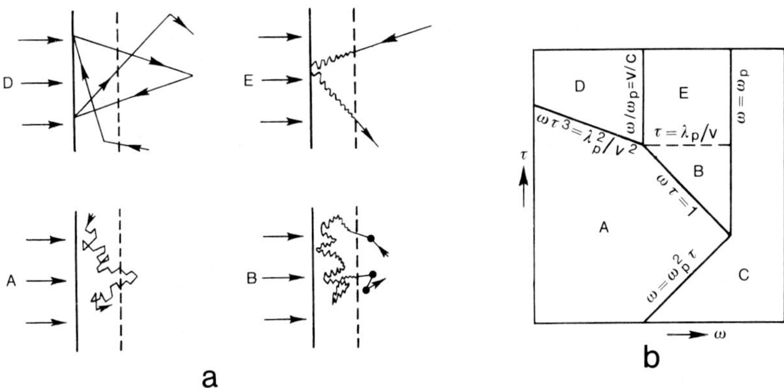

Figure 1.4 (*a*) Electron path into the skin layer. The broken line indicates the border of the skin layer while the arrows indicate the light wave incidence. The optical characteristics in the four regions depend upon the relative amplitudes of the free path of electrons, l_e, the depth of the skin layer, δ, and the mean path of an electron v_e/ω, in a $1/2\pi$ fraction of a light wave period. (A) The normal skin effect; (B) relaxation region (Drude); (D) anomalous skin effect; (E) extreme anomalous skin effect.

(*b*) The logarithmic diagram ω–τ which presents the regions illustrated above. The new region C corresponds to the transparency zone.

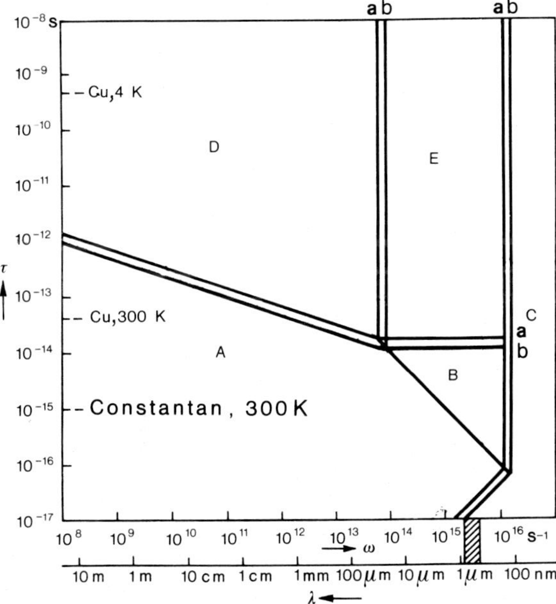

Figure 1.5 Logarithmic diagram ω–τ indicating the approximation to be used in calculating the metal absorptivity for radiation of a certain wavelength when changing the temperature.

Drude formalism is to be applied, while down at $T = 4$ K we are well into the region of the extreme anomalous skin effect (E).

In order to take into account the possibility of a change in the absorption mechanism, it is useful to introduce the expression for the metal absorptivity [26] in the form

$$A = A_D + A_A. \tag{1.54}$$

In order to give an idea of the importance of using equation (1.54) for the determination of the metal absorptivity instead of the simpler approximation $A \simeq A_D$, even if only at the room temperature, let us look at the computed $A(\lambda)$ dependence for Ag, represented in figure 1.6. One can see that for $f \simeq 0$, equation (1.54) gives a large difference as against the Drude model provisions over an extended spectral range from 3 to 300 μm while for $f \simeq 1$, the difference, even though smaller, is to be noticed within the whole generation range of the most advanced laser sources known to date.

Finally, in order to account for the interband absorption the expression (1.54) is to be completed with the corresponding value A_{IB}, determined by this effect

$$A = A_D + A_A + A_{IB}. \tag{1.55}$$

Following a suggestion in references [27, 28], the interband absorption can be accounted for within the framework of the Drude formalism—as we have already done when inferring equation (1.52)—by introducing an effective absorptivity, A'_D. We get then a different expression of the inner absorptivity of the metal, i.e. of the absorptivity which is determined by the metal properties only,

$$A = A_i = A'_D + A_A. \tag{1.56}$$

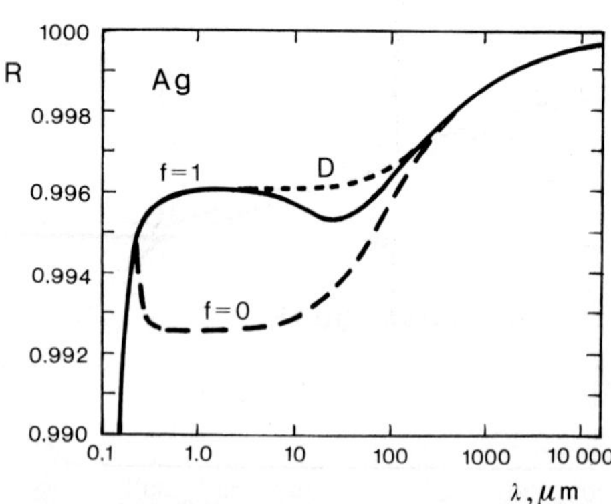

Figure 1.6 The dependence on the radiation wavelength of the reflectivity of a silver sample, computed based upon the Drude model only (full curve), and when taking into account the anomalous skin effect in two limiting situations, i.e. with $f \simeq 0$ (broken curve), and with $f \simeq 1$ (dotted curve), respectively.

Finally we must mention that, along with the introduction of A'_D, it is also useful to introduce a certain effective collision frequency

$$\Gamma_{ef} = \frac{1}{\tau_{ef}} \tag{1.57}$$

which, by analogy with equation (1.26), can be used with the approximation

$$A_i \simeq \frac{2\Gamma_{ef}}{\omega_p}. \tag{1.58}$$

If one determines experimentally the value A_i and then computes the effective collision frequency Γ_{ef} with the aid of equation (1.58), one can further evaluate, by comparison with the parameter Γ, the contribution to the absorption of the anomalous skin effect as well as of the interband transitions.

An excellent test (cf, for example, references [29–31]) for establishing the limiting radiation wavelength, λ_{IB}, below which the influence of the interband absorption becomes significant, is based upon the examination of the experimentally determined curve $\tau_{ef}(\lambda)$. A typical example of such a curve is represented in figure 1.7. The value λ_{IB} is determined as corresponding to the radiation frequency for which a sharp rise in τ_{ef} begins. The Drude theory can be applied to calculate the optical properties of metals only for $\lambda > \lambda_{IB}$.

In the case of silver, we have $\lambda_{IB} \sim 1$ μm—i.e. even for the neodymium laser ($\lambda \simeq 1.06$ μm) one can neglect the influence of the interband transitions. However, the situation can be different when irradiating other metals. For example in the case of aluminium, at $\lambda \simeq 1.06$ μm, one has $A_{IB} \simeq 4.5 \times 10^{-2}$, which is four times larger than $A_D \simeq 1.1 \times 10^{-2}$.

1.1.6. *Oblique incidence*

Whereas at perpendicular incidence the absorption of radiation does not depend on polarisation, for oblique incidence the polarisation becomes

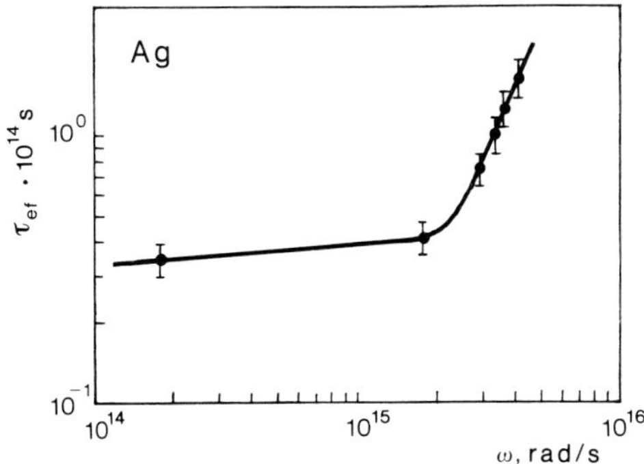

Figure 1.7 The evolution of the effective collision time, τ_{ef}, for silver, for different values of the radiation frequency.

important. For example, according to reference [32], at an incidence angle θ, the reflectivity at the metal surface of linearly polarised radiation directed parallel (index $\|$) or perpendicular ($\perp$) to the surface are given, assuming $\theta \leqslant 90°$ and

$$n^2 + k^2 \gg 1 \tag{1.59}$$

by the relations

$$R_{\|}(\theta) = \frac{(n^2 + k^2)\cos^2\theta - 2n\cos\theta + 1}{(n^2 + k^2)\cos^2\theta + 2n\cos\theta + 1} \tag{1.60}$$

and

$$R_{\perp}(\theta) = \frac{(n^2 + k^2) - 2n\cos\theta + \cos^2\theta}{(n^2 + k^2) + 2n\cos\theta + \cos^2\theta}. \tag{1.61}$$

The meaning of the indexes n and k was introduced by the expression (1.17), and further explained in the relations (1.21)–(1.23).

For the great majority of metals one has $n \gg 1$, $k \gg 1$, so that the relations (1.60) and (1.61) can be simplified

$$R_{\|}(\theta) \simeq 1 - \frac{4n}{(n^2 + k^2)\cos\theta} \tag{1.62}$$

and

$$R_{\perp}(\theta) \simeq 1 - \frac{4n\cos\theta}{n^2 + k^2}. \tag{1.63}$$

We note that under the same assumption, the relation (1.18) giving the absorptivity at perpendicular incidence takes the much simpler form

$$A(\theta = 0) = 1 - R(\theta = 0) \simeq \frac{4n}{n^2 + k^2}. \tag{1.64}$$

Relations (1.62) and (1.63) can be further transformed to obtain the corresponding expressions for absorptivity

$$A_{\|}(\theta) \simeq \frac{A(\theta = 0)}{\cos\theta} \tag{1.65}$$

and

$$A_{\perp}(\theta) \simeq A(\theta = 0)\cos\theta. \tag{1.66}$$

One is to note that the modification of A and R by oblique incidence of radiation is always accompanied by a change in the irradiation spot area, $S_s \sim 1/\cos\theta$. Therefore, for fixed parameters of the radiation incident on a sample, the intensity falling on the irradiation spot is diminished by a factor $1/\cos\theta$.

1.1.7. *The parameters n, k,* $-\varepsilon_1$, ε_2, ω_p and τ

The optical properties of metals have been studied by many authors. A standard reference is the book by Hass and Hadley [33], which, however, includes only data obtained up until 1967. A recent, more complete, reference which covers the spectral ranges of near-infrared and medium-infrared up to 12 μm is by Weaver *et al* [34]. The optical properties of Al, Co, Cu, Au, Fe, Ni, Pb, Pt, Ag, Ti and W are discussed in reference [35].

Based on the information in these works, the values of the parameters $-\varepsilon_1$, ε_2, n and k were interpolated, for $\lambda \simeq 10.6$ μm, and are presented in table 1.1.

The values of the parameters ω_p and τ, required in the evaluation of the optical properties of metals, particularly of their absorptivity, are presented in table 1.2.

The ω_p values were obtained in two ways—using equation (1.13) and by the best fitting of the data in the IR range [35]. Similarly, the τ values were either calculated according to equations (1.1) and (1.9), or obtained by a fitting in the infrared range [35]. Also the following approximate equality was used, which is fulfilled in the infrared spectral range

$$-\varepsilon_1(\omega=\Gamma)\simeq\varepsilon_2(\omega=\Gamma)\simeq\frac{\omega_p^2\tau^2}{2}. \tag{1.67}$$

This comes from the relation (valid for $-\varepsilon_1 \gg 1$)

$$-\varepsilon_1\simeq\frac{\omega_p^2}{\omega^2+\Gamma^2} \tag{1.68}$$

and which also provides for an estimation of τ.

Table 1.2 The parameters ω_p and τ at room temperature.

Metal	ω_p (10^{16} rad s^{-1})		τ (10^{-14} s)		
	From equation (1.13)	IR fitting [35]	From equations (1.1) and (1.9)	IR fitting [35]	From $-\varepsilon_1(\omega)\simeq\varepsilon_2(\omega)$
Ag	1.43	1.37	3.49	3.66	3.46
Al	2.23	2.24	0.801	0.820	0.76
Au	1.32	1.37	2.47	2.46	2.46
Cu	1.17	1.20	1.25	1.91	2.08
Pb	1.12	1.17	0.322	0.366	0.343
W	0.869	0.912	1.29	1.23	1.23

Table 1.2 does not give data for transition metals, with the exception of W, due to the difficulties in evaluating the effects of the interband transitions.

1.1.8. The applicability of the Drude model in describing metal absorptivity at $\lambda = 10.6$ μm

In this section we shall introduce a few numerical estimations of the electrical and optical properties of some metals for one particular case: at room temperature and for CO_2 laser radiation.

In the case of silver for example, the condition (1.24) is fulfilled over a wide range of wavelengths, $0.138\ \mu m \leqslant \lambda \leqslant 71.5\ \mu m$—thus including radiation of $\lambda = 10.6\ \mu m$. Using now equation (1.26), for $n_e = 5.8 \times 10^{22}$ cm^{-3}, $1/\sigma_0 = 1.61 \times 10^{-6}\ \Omega$ cm $= 1.79 \times 10^{-18}$ rad s^{-1}, $\omega_p = 1.37 \times 10^{16}$ rad s^{-1}, we obtain $A_D(293\ K) = 3.86 \times 10^{-3}$. In a similar manner one gets, for copper, $A_D(293\ K) = 4.94 \times 10^{-3}$.

As it was indicated, formula (1.26) cannot be used in all cases, as the restriction $\omega^2 \gg \Gamma^2$ is not valid for all metals. For example, in the case of Al at room temperature, with $1/\sigma_0 = 2.83 \times 10^{-6}\ \Omega$ cm $= 3.14 \times 10^{-18}$ rad s^{-1}, $n_e = 1.56 \times 10^{23}$ cm^{-3}, an electronic relaxation frequency $\Gamma = 1.25 \times 10^{14}$ rad s^{-1} is obtained, which is very close to the angular frequency corresponding to the radiation of CO_2 lasers, of $\omega = 1.78 \times 10^{14}$ rad s^{-1}. The Drude reflectivity/absorptivity can be calculated using equations (1.32),

Table 1.3 Absorptivity of metals at room temperature ($\lambda = 10.6\ \mu m$).

Metal	$A_D \times 10^{-3}$	$\langle v_F \rangle$ (10^8 cm s^{-1})	$A_A \times 10^{-3}$	$A_{min} \times 10^{-3}$
Ag	3.86	1.39	0.19	4.7 [36] 5.0 [23, 37] 7.72 [38]
Al	10.6	2.02	0.4	10.0 [39] 12.4 [14] 13.0 [40] 18.2 [41]
Au	5.9	1.39	0.19	6.0 [23, 40] 6.1 [36] 8.15 [38]
Cu	4.94	1.56	0.24	6.5 [23, 38] 7.5 [31] 7.7 [39] 11.0 [40] 11.4 [41]
Pb	44.2			
W	17.2			

(1.33) or (1.36). For example, with equation (1.33) and using data from table 1.2, we obtain $A_D(293\ \text{K}) \simeq 1.06 \times 10^{-2}$ for Al, $A_D(293\ \text{K}) \simeq 5.9 \times 10^{-3}$ for Au, $A_D(293\ \text{K}) \simeq 4.42 \times 10^{-2}$ for Pb and $A_D(293\ \text{K}) \simeq 1.72 \times 10^{-2}$ for W.

The A_D values computed at room temperature, the Fermi velocity $\langle v_F \rangle$, and the A_A values for $f = 1$ are collected in table 1.3 for several metals, together with the smallest experimentally determined absorptivities, A_{min}, of the respective metals. We notice the rather good agreement between the computed values of the intrinsic absorptivity of these metals, $A_i = A_D + A_A$, and the experimental data. The differences yet to be observed point to the necessity of taking into account, in the respective cases, the interband absorption. For example, according to reference [23] in the case of aluminium, we have $A_{IB} \simeq 2 \times 10^{-3}$.

1.2. Real metallic surfaces

The mere inspection of the data available from the literature concerning the experimentally determined values of the absorptivity of metals reveals a large dispersion of data—even when referring to one and the same metal—and most often quite a large disagreement with the provisions of the previously examined models. This disparity mainly originates in the fact that, with many of the various experiments, the quality of the surface of the irradiated samples departs from what is normally understood by an ideal one. The experimentally measured absorptivities were not therefore determined only by the intrinsic properties of the metals—and consequently could not possibly be accurately described by the formula $A_i = A_D + A_A + A_{IB}$—but also, and often to a large extent, by the optical properties of the sample surface, which contributes a supplementary absorptivity, A_{ext}, i.e.

$$A = A_i + A_{ext}. \tag{1.69}$$

Accordingly, A_{ext} is called the external absorptivity of the sample.

The A_{ext} value is in turn determined by the supplementary absorption of radiation by surface roughness (A_r), the influence upon absorptivity of various defects and impurities (A_{id}), of the oxide layers and of the layers consisting of adsorbed substances (A_{ox}), so that

$$A_{ext} = A_r + A_{id} + A_{ox}. \tag{1.70}$$

We shall examine separately all these factors.

1.2.1. The surface roughness

The influence of surface roughness on absorptivity can manifest itself in two different ways. There is first the absorption increase on surface areas where the radiation incidence is $\theta \neq 0$ and also in grooves and cracks, which favour a waveguided propagation of radiation. Secondly, there are the collective effects, related to the generation of surface electromagnetic waves on the periodical surface microrelief. The limiting case of such a relief—the resonant surface periodical structures (having a period of the order of magnitude of the incident radiation wavelength)—will be examined in Chapter 5. We shall give here an indication of the approach to the calculation of A_r for a surface with random roughness.

Various correlative aspects of this topic, of consequence in both theory and practice, were investigated among others in references [42–50]. Generally, the standard case for investigation was a metallic sample with highly processed surfaces, characterised by a roughness δ_σ, much smaller than the wavelength of the incident laser radiation, i.e. $\delta_\sigma/\lambda \ll 1$. In such a case the calculation of A_r can be conducted in first-order perturbation theory with respect to A_i.

A simple estimation of the reflectivity for a rough surface, R_r, as compared with the reflectivity of a very smooth surface of the same material, $R \approx R_i$, can be obtained with the expression [24]

$$R_r \simeq R_i \exp[-(4\pi\delta_\sigma/\lambda)^2]. \tag{1.71}$$

As equation (1.71) was established on the assumption of a Gaussian distribution of the roughness height on the sample surface, there may be erroneous evaluations in the many circumstances when this hypothesis does not stand. In reference [51], the following expression for A_r was established, which results from a surface roughness described by the distribution function $\zeta(x, y)$:

$$A_r = \frac{\varepsilon_2|1-\varepsilon_c|^2 q_0}{2\pi^2} \times \int \mathrm{d}^2k \frac{|\zeta(\boldsymbol{k})|^2}{S_s} \frac{1}{\mathrm{Im}(q')} \left(\frac{k^2+|q'|^2}{|q'+q\varepsilon_c|^2} \left| \frac{q_0' q \cos\Phi}{q_0\varepsilon_c + q_0'} + \frac{(\omega/c_0) q \sin\Phi}{q_0+q_0'} \right|^2 + \frac{(\omega/c_0)^4}{|q+q'|^2} \left| \frac{(\omega/c_0)\cos\Phi}{q_0+q_0'} + \frac{q_0' \sin\Phi}{q_0\varepsilon_c + q_0'} \right|^2 \right). \tag{1.72}$$

Here $\boldsymbol{k} = k(\cos\Phi \boldsymbol{i} + \sin\Phi \boldsymbol{j})$, where $\boldsymbol{i}, \boldsymbol{j}$ are the unit vectors of the axes $0x$, $0y$, of the incidence plane, and $q_0 = \omega/c_0'$, $q_0' = q_0\sqrt{\varepsilon_c}$, $q = [(\omega/c_0)^2 - k^2]^{1/2}$, $q' = [\varepsilon_c(\omega/c_0)^2 - k^2]^{1/2}$ and $\delta_\sigma = (\langle\zeta^2(x, y)\rangle)^{1/2}$. The parentheses $\langle\ \rangle$ indicate an averaging over space. The function under the integral in equation (1.72) contains the Fourier transform of the roughness profile, as given by

$$\zeta(\boldsymbol{k}) = \int \mathrm{d}^2\rho\,\zeta(\boldsymbol{\rho}) \exp(\mathrm{i}\boldsymbol{k\rho}) \tag{1.73}$$

where $\boldsymbol{\rho} = (x, y)$ and ε_2, ε_c, ω, c, have the same meaning as above.

For a random distribution of the roughness of the sample surface, a value averaged over the entire sample of area S_s is used in equation (1.72)

$$\frac{\langle \zeta(\boldsymbol{k}) \rangle}{S_s} = \int d^2 u G(u) \exp(i\boldsymbol{k}\boldsymbol{u}) \equiv g(\boldsymbol{k}) \tag{1.74}$$

which, with

$$G(u) = \delta_s^2 \exp[-(u/\sigma_s)^2] + \frac{\delta_e^2}{1+(u/\sigma_1)^2} \tag{1.75}$$

becomes

$$g(\boldsymbol{k}) = \pi(\delta_s \sigma_s)^2 \exp(-|\boldsymbol{k}|^2 \sigma_s^2/4) + 2\pi(\delta_1 \sigma_1)^2 K_0^m(|\boldsymbol{k}|\sigma_1). \tag{1.76}$$

Here the average roughness δ_s, δ_l, as well as the corresponding distances σ_s, σ_l characterise the space correlation over short and long distances respectively across the surface [42, 50], while K_0^m is the modified Bessel function, of second kind and of zero order.

When working with a set of roughness parameters corresponding to a high-quality metal mirror

$$\delta_s = 3\ \text{Å} \qquad \sigma_s = 0.15\ \mu\text{m} \qquad \delta_l = 30\ \text{Å} \qquad \sigma_l = 2\ \mu\text{m} \tag{1.77}$$

and with samples made of Cu (findings also valid for Au, Ag and Al), equations (1.72)–(1.77) show that the additional absorptivity induced by the excess roughness does not exceed more than 1–2% (figure 1.8) of the intrinsic

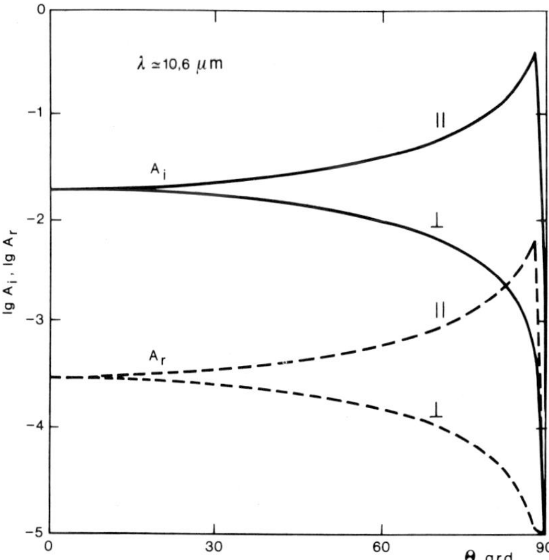

Figure 1.8 The dependence of absorptivity on the angle of incidence, θ. A_i stands for the intrinsic absorptivity characteristic of the sample material while A_r is the contribution to absorptivity due to the roughness of the surface.

absorptivity of the sample as established by the formula

$$A_i = \frac{2q_0\varepsilon_2}{Im(q_0')}\left(\frac{|q_0'|^2}{|q_0' + \varepsilon_c q_0|} + \frac{(\omega/c_0)^2}{|q_0 + q_0'|^2}\right). \qquad (1.78)$$

However A_r rapidly increases when δ_s increases. Consequently, while the term A_r can be neglected when calculating the absorptivity of high-quality mirrors, it may, however, become significant in the case of conventional metal samples.

For metal samples characterised by space structures which are either periodic or quasi-periodic, the ratio A_r/A_i substantially increases so that it can exceed unity. Its value would markedly depend on both the polarisation and the incidence angle of the radiation. Assuming that over the entire irradiated area the surface is ideally plane along one direction while also exhibiting a roughness (possibly a periodical roughness), the following expression has been established [48] for the supplementary absorptivity, following from equation (1.72):

$$A_r = \varepsilon_2 \frac{\omega}{c_0} |\sqrt{\varepsilon_c} - 1|^2 \sum_k |\zeta(k)|^2 \frac{(|q'|^2 + k^2)}{\text{Im}\ (q')} \frac{|q|^2}{|q' + \varepsilon_c q|^2}. \qquad (1.79)$$

The following expressions for A_r, with an incident radiation polarised linearly, either parallel, $A_r^{\parallel}$, or perpendicular, $A_r^{\perp}$, to the incidence plane were also established [48]:

$$A_r^{\parallel} = 2\varepsilon_2 \frac{\omega}{c_0} |\sqrt{\varepsilon_c} - 1|^2 \sum_k |\zeta(k)|^2 \frac{(|q'|^2 + k^2)}{\text{Im}(q')} \frac{|q|^2}{|q' + \varepsilon_c q|^2} \qquad (1.80)$$

and

$$A_r^{\perp} = 2\left(\frac{\omega}{c_0}\right)^5 \varepsilon_2 |\sqrt{\varepsilon_c} - 1|^2 \sum_k \frac{|\zeta(k)|^2}{\text{Im}(q')|q' + q|^2}. \qquad (1.81)$$

1.2.2. Defects and impurities

A great variety of impurities and microimpurities are scattered on and in the surface of any real metallic sample, and each can contribute to a supplementary absorption of radiation. To illustrate, let us examine the results of several investigations dealing with the microscopic analysis of typical metallic surfaces.

Among the most common surface microimpurities are dust particles of different shapes and sizes [52]. Such particles on the metal surface function as centres of sharply enhanced local absorptivity. Various methods have been proposed for the elimination of dust contamination of surfaces, ranging from blowing a very clean gas jet, to surface cleaning based on the use of solvents, or of electrostatic devices.

Other common surface impurities are the abrasive particles left behind by polishing, either on the surface or stuck into the bulk of the material (figure 1.9). According to the data in table 1.4 [53, 54], the electrochemical

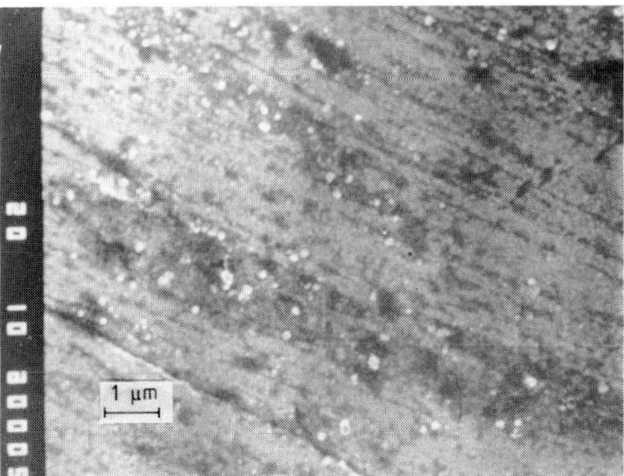

Figure 1.9 Abrasive particles (the small white spheres) remaining on the surface after mechanical polishing of a copper mirror.

polishing and the annealing at high temperatures considerably reduce the concentration, Γ_p, of such inclusions. In general, comparing data in tables 1.3 and 1.4 one can see that the abrasive particles would not significantly increase the absorptivity of the metals.†. It is also proper to emphasise the lack of any direct connection between A and Γ_p; even the logical connection $A_{id} \sim \Gamma_p$ is broken—as can be seen by comparing the data in table 1.4 referring to similar samples submitted to mechanical polishing and high-temperature annealing. Indeed, although the surface of the copper sample is less polluted with abrasive particles than the sample submitted to mechanical polishing, one can notice that the absorptivities exhibited by the two samples compare with each other inversely as expected. In the given case this behaviour could be, in the first place, an effect of a poor vacuum during annealing, resulting in the formation of an oxide layer on the surface of the sample, and so inducing a supplementary absorptivity, A_{ox}. Also the thermal processing (annealing) can introduce new defects.

Among the defects occurring in the bulk of metallic materials we mention pores, cracks (figure 1.10(*a*)) and grooves (cavities) (figure 1.10(*b*)) whose role in increasing the local absorptivity was shown in references [55–58].

An important part in increasing the absorptivity of a (laser) irradiated metallic sample can be played by defects consisting of the metal itself, which, for some reason, are thermally insulated from the rest of the material

† However, in general the absorption depends on the nature of the embedded particles. Indeed diamond has a lower absorption than copper—small diamond particles do not increase significantly the absorption of 10.6 μm radiation. The situation can be different with other polishing materials in a direct relation with their absorption at the wavelength of use: e.g. SiC absorbs in the visible and not so much in the IR, Fe_2O_3 absorbs highly at all wavelengths, etc. As a general trend we emphasise that in order to reduce/eliminate the increase of absorption by the inclusion of particles, the polishing materials should be transparent at the wavelength of use.

Table 1.4 Characteristics of copper mirrors manufactured by different surface processing methods.

Kind of processing	Γ_p (cm^{-2})	A
Polishing with an abrasive (mechanical polishing)	1×10^8	0.012 ± 0.001
Annealing	6×10^7	0.0141 ± 0.001
Electrochemical polishing	3×10^7	0.009 ± 0.001

[53, 54, 57, 59, 60]. A typical example of this kind of defect is metallic 'flakes', like those shown in figure 1.11. The cause of the enhanced absorption is in this case a combined effect of the overheating of these defects in comparison with the metallic sheet nearby, and of the temperature dependence

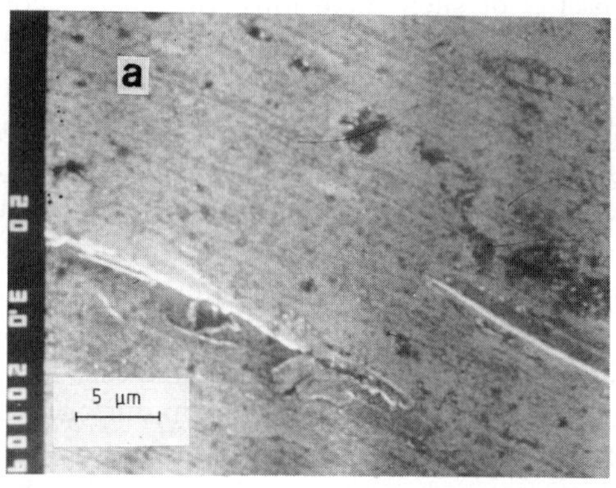

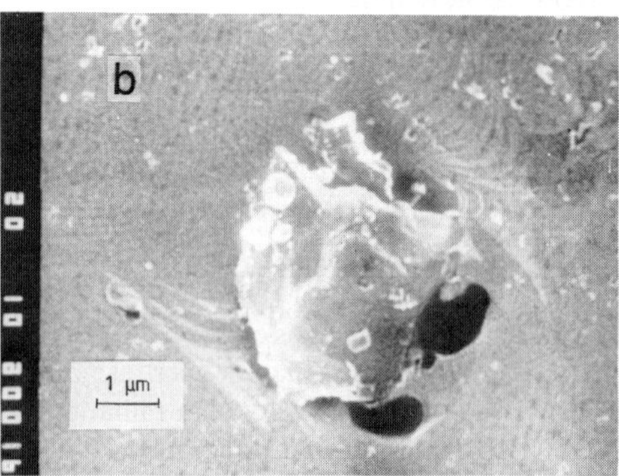

Figure 1.10 Microcracks (*a*) and a groove (microcavity) (*b*) on the surface of a copper sample.

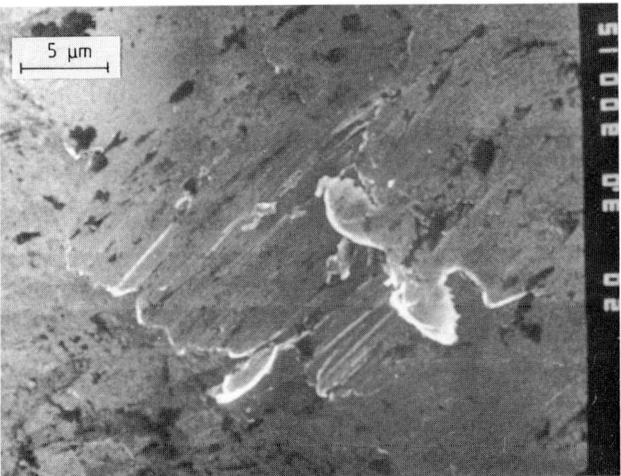

Figure 1.11 Micrograph of surface defects thermally insulated from the metallic sample. The 'flake' type.

of the metal absorptivity: indeed, 'hot' points showing high absorptivity appear against the 'cold' background of the sheet surface. Such a behaviour is stressed in the case of high-intensity (pulsed) laser irradiation of metallic surfaces.

Of course, both metallic and non-metallic defects can be overheated—in the latter case due to a poorer thermal conductivity as compared with the metal's, which limits the heat transfer, even in the case of a good contact with the substratum.

In order to evaluate the overheating ΔT one can use the well known expression of the temperature excursion of a special dielectric microsphere of radius a embedded in the metallic sheet

$$\Delta T = \frac{A_d I_0 \sqrt{\kappa \tau_p}}{k_T} 2\,\mathrm{i\,erfc}\,\frac{a}{2\sqrt{\kappa \tau_p}}. \tag{1.82}$$

Here I_0 is the intensity of the incident laser radiation, τ_p the irradiation duration, k_T and κ the thermal conductivity and the thermal diffusivity of the material and A_d the absorptivity, at the radiation absorption into the defects. The following integral notations are used

$$\mathrm{i\,erfc}\; y = \int_0^\infty \mathrm{erfc}\;\xi \; \mathrm{d}\xi \tag{1.83}$$

and

$$\mathrm{erfc}\; y = (2/\pi)^{1/2} \int_y^\infty \exp(-\xi^2)\; \mathrm{d}\xi. \tag{1.84}$$

One of the most efficient methods to be used for removing defects and impurities from strongly polluted (dirty) surface layers consists of thermal processing of the surface by the action of laser radiation in high vacuum (in order to avoid oxidation).

One can follow from figure 1.12, the heating diagram, $T(t)$ and the temperature variation of absorptivity, $A(T)$, for a copper sample when submitted in vacuum to cw irradiation by a CO_2 laser [60]. It can be seen that even for rather short-time heating, up to 750°C, laser annealing results in a twofold diminishing in $A(T)$. At the origin of this effect is the removal of microparticles and films from the surface, and the diffusion of impurities from the surface layer down into the sample (see also Chapter 6).

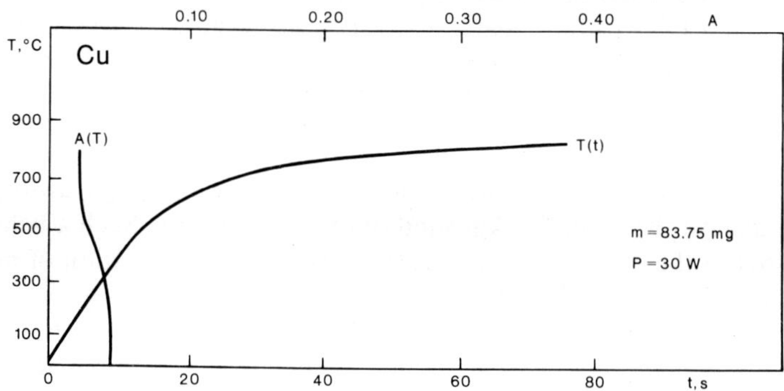

Figure 1.12 Heating diagram for a thermally insulated copper sample of mass $m = 83.7$ mg, cw irradiated in vacuum with a CO_2 laser rated at a power $P = 30$ W. Also shown is the corresponding temperature variation of the sample's absorptivity. The laser cleaning effect is clearly visible.

1.2.3. Surface-adsorbed substances

Identification of surface-adsorbed impurities by mass spectroscopic analysis of a small quantity of substance expelled as an effect of power-laser irradiation ($\lambda = 1.06$ μm and peak intensity $I_0 \leqslant 10^{13}$ W m^{-2}) was reported for the first time by Apollonov *et al* [61]. Targets made of $^{27}_{60}$Co, $^{47}_{181}$W and $^{83}_{209}$Bi were used. The signals of multiply ionised ions up to Co^{+25}, Ag^{+16}, W^{+11} and Bi^{+19} were identified together with a group of signals which seemed to correspond to lighter-weight impurity ions—such as carbon, nitrogen and oxygen. Similarly, electrostatic analysis of the substance expelled from the surface of some tungsten targets by power-laser irradiation ($\lambda = 0.6943$ μm)

[62] has led to the identification of carbon, oxygen and nitrogen ions (singly or doubly ionised). In neither case do the authors give details regarding the history, preparation and processing of the targets used in the experiments, which were characterised only as being 'fresh'.

A much more elaborate study was performed by Dinger *et al* [63, 64]: the surface impurities of tantalum, tungsten and gold samples were investigated. The samples, which were characterised as being of a technical purity grade, were subject to prior cleaning with organic solvents. Time-of-flight techniques (using a Faraday electrode and a retarding-potential analyser) were applied for the analysis of a layer of expelled substance with a thickness of only some tenths of a micrometre which was peeled off in high vacuum ($\sim 10^{-7}$ Torr) with a surface spot having a diameter of $\simeq 170\,\mu$m under the action of a laser pulse ($\lambda = 1.06\,\mu$m) of 13 mJ/35 ps. Typical forms of the ion current for the first and second radiation pulse, respectively, are given in figure 1.13.

By processing such signals it was shown that, for tantalum and tungsten targets, the impurities may amount to up to 85% of the entire quantity of

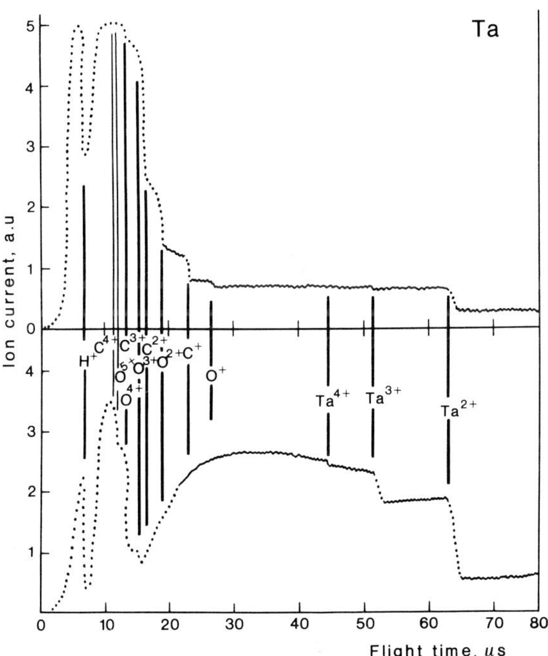

Figure 1.13 The ion current ignited by a Nd:YAG laser pulse from a tantalum target. Retarding potential $\simeq 800$ V, flight distance $\simeq 258$ cm. The upper and lower spectra refer to the first and second irradiations, respectively, on the same location on the sample surface.

expelled substance during the first irradiation pulse. Carbon, hydrogen and oxygen ions of different electric charges were identified, together with ions of the irradiated metal.

When a second laser pulse was applied after 5 s on the same irradiation spot, the situation changed and the impurities represented no more than 15% of the total amount of expelled substance, which was markedly diminished as compared with the first pulse. The metal ions dominating the signal presented in this case a higher kinetic energy as well as a higher degree of ionisation.

Surface contamination of gold proved to be, as expected, considerably lower. The impurities represented only $\simeq 3\%$ of the mass expelled from the sample surface.

The origin of such pollution on the surfaces of real samples is mainly the adsorption onto the metal surface of various organic substances. The difference between the amount of impurity material on the surfaces of gold, in comparison with the other investigated metals, points directly to the important role played in the formation of impurity layers by chemisorption of gases on surfaces.

When further providing for the additional detection of neutral species with the help of an ionising agent, these investigations were improved [65, 66] up to the level of developing and implementing a new experimental technique called laser-induced desorption, which enables the identification and elimination, with a high spatial resolution, of the surface contaminants and avoids melting or damaging of the surface. The preferential adsorption of contaminants on surface defects (scratches, cracks, pits, pores, etc) was thereby demonstrated, though no significant desorption of water from the irradiated mirrors was noted.

It is to be emphasised that the light elements adsorbed on surfaces—shown in all these studies—can cause a significant supplementary absorption of radiation, either directly or by the formation of different chemical compounds (especially organic compounds).

1.2.4. Oxides

When kept in air, the surfaces of conventional metals are most often covered by an oxide layer that sometimes shows a multilayer structure.

The thickness, x, and structure of the oxide layer depend on the preparation and history of the metal sample. On the other hand, the rather thick oxide layers can cause an increase in the sample absorptivity, by as much as an order of magnitude and more [53, 54, 59] (for details see Chapter 6).

Of course, the influence of the oxide layers differs for radiation in different spectral ranges. To give an example, let us consider oxidised aluminium. Under normal conditions, the natural oxide layer on aluminium remains rather thin ($x < 100$ Å). However, even a layer of low x may have as an effect the fulfilment of the condition $A_{ox} \geqslant A_i$ in the case of UV radiation generated

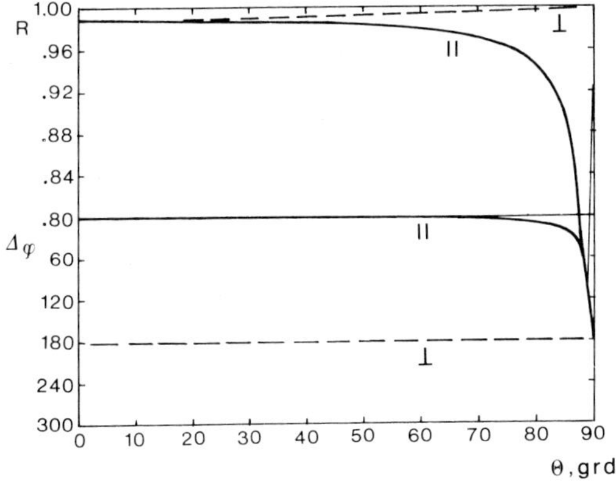

Figure 1.14 The reflectivity amplitude, R, and the phaseshift, $\Delta\varphi$, as functions of the radiation ($\lambda = 10.6\ \mu$m) incidence angle, θ, on the surface of an aluminium mirror coated with an oxide layer $\simeq 40$ Å thick.

by excimer laser sources with $\lambda \leqslant 2200$ Å [40, 67, 68]. The same Al_2O_3 films are, however, completely transparent to the radiation of CO_2 lasers.

The reflectivity of an aluminium optical surface coated with a natural oxide layer of thickness $x = 40$ Å (a typical value for the final thickness of the oxide forming on aluminium samples kept in air) was calculated [67] for a wide range of wavelengths from ultraviolet to mid infrared. An example corresponding to the CO_2 laser radiation is presented in figure 1.14, where it is possible to follow the dependence on the value of radiation incidence angle, θ, of both the reflectivity amplitude, R, and the phaseshift, $\Delta\varphi$, appearing as a result of reflection on the oxide–metal system of the radiation linearly polarised either parallel or perpendicular to the incidence plane. The comparison of the curves $R(\theta)$ with the computed $R_i(\theta)$ curves for aluminium show that in this case $A_{ox} \leqslant 0.016\ A_i$, i.e. $A_{ox} \ll A_i$.

We emphasise the absence of any influence of the oxidation upon the optical properties of aluminium in the mid infrared range.

1.3. Temperature variation of the absorptivity of metals

For a great many practical evaluations related to the use of laser techniques, reliable information is necessary concerning the behaviour of metal absorptivity/reflectivity as a function of temperature.

The temperature dependence of the optical properties of metals has been investigated theoretically in many papers (see, for example, references [15, 30, 58, 69, 70]). To begin with, let us discuss this problem in the simplest case—the approximation of the Drude model.

From equation (1.9), taking into account the known (tabulated) temperature dependence of the DC electric conductivity, $\sigma_0(T)$, it results that the electron relaxation frequency increases with the temperature increase. Furthermore, from equations (1.9) and (1.21)–(1.23) one obtains

$$\frac{\mathrm{d}n}{\mathrm{d}\Gamma}=\frac{\Gamma}{2n}\frac{\omega_p^2}{(\omega^2+\Gamma^2)^2}\left(1+\frac{2-\omega_p^2/\omega^2}{2[1-\omega_p^2(2-\omega_p^2/\omega^2)(\omega^2+\Gamma^2)^{-1}]^{1/2}}\right). \tag{1.85}$$

When $\omega_p^2/\omega^2>4$, this derivative changes sign from positive to negative, for

$$\Gamma^2(T)=\frac{3\omega_p^2/\omega^2-4}{\omega_p^2/\omega^2-4}\,\omega^2. \tag{1.86}$$

Accordingly, n would increase with the temperature up to a maximum value, then it would decrease.

Similarly, for the absorption index k, from (1.9) and (1.21)–(1.23), it results

$$\frac{\mathrm{d}k}{\mathrm{d}\Gamma}=-\frac{\Gamma}{2k}\frac{\omega_p^2}{(\omega^2+\Gamma^2)^2}\left(\frac{2-\omega_p^2/\omega^2}{2[1-\omega_p^2(2-\omega_p^2/\omega^2)(\omega^2+\Gamma^2)^{-1}]^{1/2}}-1\right). \tag{1.87}$$

From equation (1.87) one notices that whenever $\omega_p^2/\omega^2>4$, the derivative $\mathrm{d}k/\mathrm{d}\Gamma$ is always negative, so that k diminishes with the temperature increase, as we have $\mathrm{d}\Gamma/\mathrm{d}T>0$. Consequently, the depth of the skin layer as given by the relation

$$\delta_D\simeq\frac{c_0}{\omega k} \tag{1.88}$$

increases with temperature. Here c_0 stands for the speed of light.

And, finally, from equations (1.9) and (1.15)–(1.19) it results

$$\frac{\mathrm{d}R}{\mathrm{d}\Gamma}=-\frac{(\omega_p^2/\omega^2)(|\varepsilon_c|+1)}{\sqrt{2}|\varepsilon_c|[|\varepsilon_c|+1+\sqrt{2}(|\varepsilon_c|+\varepsilon_1)^{1/2}]^2(|\varepsilon_c|+\varepsilon_1)^{1/2}}\frac{\mathrm{d}\varepsilon_1}{\mathrm{d}\Gamma}. \tag{1.89}$$

The derivative

$$\frac{\mathrm{d}\varepsilon_1}{\mathrm{d}\Gamma}=\frac{2\omega^2\Gamma}{(\omega^2+\Gamma^2)^2} \tag{1.90}$$

is positive. Consequently

$$\frac{\mathrm{d}R}{\mathrm{d}\Gamma}<0 \qquad \text{and} \qquad \frac{\mathrm{d}R}{\mathrm{d}T}<0 \tag{1.91}$$

i.e. the reflectivity of metal samples decreases, while their absorptivity increases with temperature.

This behaviour of the functions $A(T)$ and $R(T)$ has been confirmed by most of the experimental studies. Nevertheless it has to be mentioned that when the interband absorption becomes significant it may result in unexpected situations, in which $\mathrm{d}R/\mathrm{d}T=0$ or even $\mathrm{d}R/\mathrm{d}T>0$. A case of this kind is noted, for example in [30], regarding the behaviour of the reflectivity of some vacuum-deposited aluminium films. Also the sample surface roughness can play an important role in the determination of the function $A(T)$.

We shall examine the concrete form of the dependence $A(T)$ for ideally flat surfaces in two different temperature ranges: from room temperature T_0 up to the melting point, T_m; and for $T>T_m$.

1.3.1. Solid-phase metal

We shall start with the Drude model approximation. In this case the quantity $A_i=A_D$ is to be obtained with equation (1.26) whenever the condition (1.24) is fulfilled, and is proportional to the electron relaxation frequency Γ. Assuming the absence of any defect and impurity ($\Gamma_{ed}=0$) and neglecting electron–electron collisions, we obtain $\Gamma \simeq \Gamma_{ep}$.

The electron–phonon collision frequency Γ_{ep} is approximately equal to the value Γ_{ep}^{cl} of the classical frequency determined by the DC metal resistivity

$$\Gamma_{ep}^{cl}=\frac{n_e e^2 r_0(\omega=0)}{m^*} \tag{1.92}$$

where $r_0=1/\sigma_0$ is the resistivity of the pure metal.

The values r_0 $(\omega=0)$ are tabulated for different temperatures and they increase with increase in temperature. Correspondingly, the collision frequency also increases; its temperature dependence for the particular case of aluminium can be presented in the form [15]

$$\Gamma_{ep}(T)=\frac{n_e e^2}{m^*}\left(1.22\,r_0(T_0)+\frac{r_0(T_m)-1.22\,r_0(T_0)}{T_m-T_0}(T-T_0)\right). \tag{1.93}$$

As a result, taking into account the following relation (obtainable from equation (1.26))

$$A(T)=A(T_0)\frac{r_0(T)}{r_0(T_0)} \tag{1.94}$$

one may conclude that the metal absorptivity depends linearly on temperature.

However, the experimental observation of the dependence (1.94) is particularly difficult in the case of ultra-pure metals, even with the help of the best available technology, as one has to work with perfect metallic layers, free of any defect or impurity, and with a perfectly clean surface. That is why in practice there is open disagreement between various experimental determinations of the $A(T)$ dependence. However, most of the available data on the temperature dependence of metals absorptivity—either obtained experimentally or calculated theoretically—may generally be described by a rather simple function of the type

$$A(T)=A_0+A_1T. \tag{1.95}$$

This dependence was directly inferred by fitting the data available in the infrared range with a general polynomial

$$A(T)=\sum_i A_iT^i \qquad i=0,1,2,\ldots \tag{1.96}$$

where T is to be introduced in kelvins.

As one can see from the results obtained on high-purity metals given in table 1.5, the formula (1.95) works well—as generally A_2 is practically zero or very small—whenever there is no influence on absorption by interband transitions (e.g. for radiation with $\lambda=10.6\ \mu$m). In table 1.5 are also given the values of the parameters ω_p and $\tau=1/\Gamma$ where the approximation $\Gamma=\Gamma_{ep}$ is applied—as used in reference [41], and computed by two different methods.

However, the reliability and accuracy of the theoretical results are still insufficient, mainly as a result of the difficulty of correctly choosing the input data necessary when evaluating the fundamental parameters. For example, the concentration of the conduction electrons—which in turn determines the values of ω_p and Γ—is to be obtained from

$$n_e=n_0n_a \tag{1.97}$$

where n_a stands for the concentration of metal atoms while n_0 is the number of conduction electrons per metal atom. While the situation is rather clear for n_a, in different papers one would find quite different values of n_0. For example, in the case of Al, $n_a=6\times10^{22}\ \mathrm{cm}^{-3}$, and according to the experiments reported in reference [23], on annealed and non-annealed vacuum-deposited aluminium films, the values $n_0=1.1$ and $n_0=1.3$ were obtained. On the other hand, $n_0=3$ is used in reference [23].

And, finally, when the metal surface is far different from an ideal one—as happens in most cases—the temperature modification of its optical properties can show particular features which have to be taken into account for the sake of general accuracy. In figure 1.15 the evolutions $A(T)$ are given, as obtained experimentally [15] in the case of CO_2 laser heating of some aluminium samples which were mechanically processed (unpolished) and then cleaned with various solutions. Curve (1) corresponds to the laser heating of the base material while curves (2) and (3) correspond to the repeated heating of the same sample. First we notice that the initial cold absorptivity of the metal surface is $A(T_0)\simeq0.04$, significantly exceeding the intrinsic absorptivity of the pure metal (see table 1.3). Secondly, we observe that with every subsequent heating of the metal sample, the $A(T_0)$ value diminishes along with the

Table 1.5 The temperature dependence of absorptivity of metals in solid (s) and liquid (l) phases.

Metal	ω_p (10^{16} rad s^{-1})	τ (10^{-14} s)	Phase	A_0	A_1	A_2	$A_1^{exp} \times 10^{-5}$
Ag	1.43	3.49	s	−6.88(−5)	1.59(−5)		1.6 [41, 71]
			l	1.30(−2)	1.64(−5)		
Ag	1.37	3.66	s	−7.03(−5)	1.59(−5)		2 ± 0.7 [39]
			l	1.30(−2)	1.66(−5)		
Al	2.23	0.801	s	1.47(−3)	3.20(−5)		3.2 [41]
			l	2.91(−2)	1.76(−5)		
Al	2.24	0.821	s	1.34(−3)	3.13(−5)		3 ± 0.5 [39]
			l	2.88(−2)	1.70(−5)		
Au	1.37	2.47	s	9.43(−4)	2.34(−5)		
			l	3.46(−2)	1.48(−5)		
Au	1.32	2.46	s	9.54(−4)	2.36(−5)		
			l	3.46(−2)	1.50(−5)		
Cu	1.17	1.25	s	1.92(−3)	3.88(−5)		1.4 ± 0.5 [39]
			s				1.5 ± 0.5 [72]
Cu	1.20	1.91	s	−2.83(−5)	2.59(−5)		1.56 ± 0.27 [71]
							2.6 ± 3.9 [41]
Pb	1.12	0.322	s	7.16(−3)	1.25(−4)		
			l	9.93(−2)	2.47(−5)		
Pb	1.17	0.366	s	4.67(−3)	1.14(−4)		
			l	8.88(−2)	2.25(−5)		
W	0.869	1.29	s	−2.50(−3)	7.29(−5)	−9.25(−9)	
W	0.912	1.23	s	−2.10(−3)	7.16(−5)	−9.09(−9)	

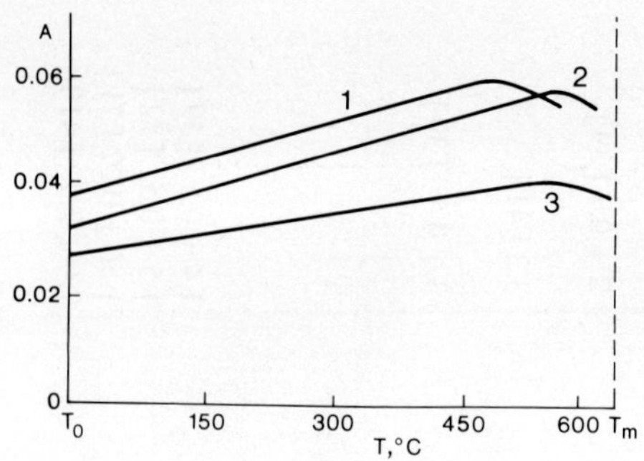

Figure 1.15 Experimental curves of the temperature dependence of the absorptivity of an aluminium sample, at different values of the 'cold' absorptivity $A(T_0)$. See text for explanation of curve labels.

cleaning of the metal surface. Thirdly, the dependence $A(T)$ keeps in all cases close to the linear one—even though the curve slope A_1 differs from the predictions of the Drude model and depends on $A(T_0)$.

Along with the influence of the surface microrelief and of the surface absorption centres—through the intermediary of the term A_{id}—the existence of defects inside the metal itself would also lead to an increase in the frequency of electron collisions. The introduction of a total collision frequency was accordingly proposed in reference [15] in the form $\Gamma = \Gamma_{ep} + \Gamma_{eff}$, where Γ_{eff} stands for an effective collision frequency, not depending on temperature and taking into account the electron scattering not only on the defects into the skin layer, but also the supplementary absorptivity related to the surface roughness. Analysis of the results obtained showed, however, that this approximation is not entirely correct, and one has to consider a temperature dependence of Γ_{eff}.

1.3.2. Liquid-phase metal

At the transition from the solid (index 's') to the liquid ('l') phase, the number of conduction electrons per one metal atom, n_0, the metal density—determining n_a—and the DC resistivity of the metal, change simultaneously. For example for Al we have $n_0^l = 3$, $n_a^l = 0.87n_a^s$, $r_0^l = 1.82r_0^s$. Consequently we may expect a stepwise increase in the absorptivity of the metal as a result of the transition from the solid phase to the liquid phase at the melting point of the metal. The computed values of the absorptivity of a series of metals for CO_2 laser radiation are given in table 1.6, at the melting point, $T = T_m$, in solid phase, $A^s(T_m)$, and in liquid phase, $A^l(T_m)$, respectively (as in table 1.2, the data in table 1.6 were obtained under two approximations, i.e. the Drude model and a fitting over the IR range, respectively). From table 1.6 one can clearly see that the phase transformation

Table 1.6 The absorptivity of $\lambda \simeq 10.6$ μm radiation by some metals at the melting point, in the solid phase and in the liquid phase respectively, as obtained with the aid of Drude's formalism or from the best fitting over the IR range.

Metal	T_m (K)	$A^s(T_m) \times 10^{-2}$		$A^l(T_m) \times 10^{-2}$	
		A^s_D	A^s_{IR}	A^l_D	A^l_{IR}
Ag	1234	1.97	1.95	3.74	3.35
Al	933	3.01	3.05	5.72	4.47
Au	1337	2.85	3.25	5.42	5.46
Cu	1356	2.78	3.51	5.92	
Pb	601		7.31		10.2

at the melting point is accompanied by an amplification by a factor of 1.5–2 in the absorptivity.

The computed values of the ratio $A^l(T_m)/A^s(T_m)$ are in rather good qualitative agreement with the few experimental data available. For example, in papers [73–75] the temperature evolution was recorded of the absorptivity for CO_2 laser radiation of some single-crystal Al samples with a non-polished surface (therefore $A(T_0) > A_i(T_0)$). Some results of these measurements are reproduced in figure 1.16. We observe that the absorptivity increases almost 1.3 times at $T \simeq T_m$ by the metal transition from a solid to a liquid phase—a value which is, however, slightly lower than the computed value for a clean,

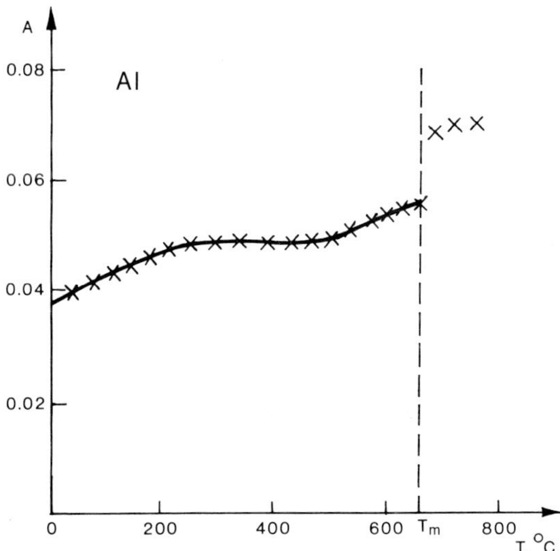

Figure 1.16 Temperature dependence of the absorptivity, $A(T)$, of an unpolished sample prepared from single-crystal aluminium. One can see the stepwise increase in A at the melting temperature, T_m.

ideally plane surface. We also emphasise the quite unexpected form of the $A(T)$ curve at $T \leqslant T_m$.

The absorptivity continues to increase with the temperature increase for $T > T_m$, as one can see from the computed curves given in figure 1.17, and can be approximated as in the solid phase by expressions of type (1.95) or (1.96), where A_0 is then the liquid metal absorptivity at the melting point. The values A_0 and A_1 corresponding to the computed $A^l(T)$ curves in figure 1.17 are given in table 1.5.

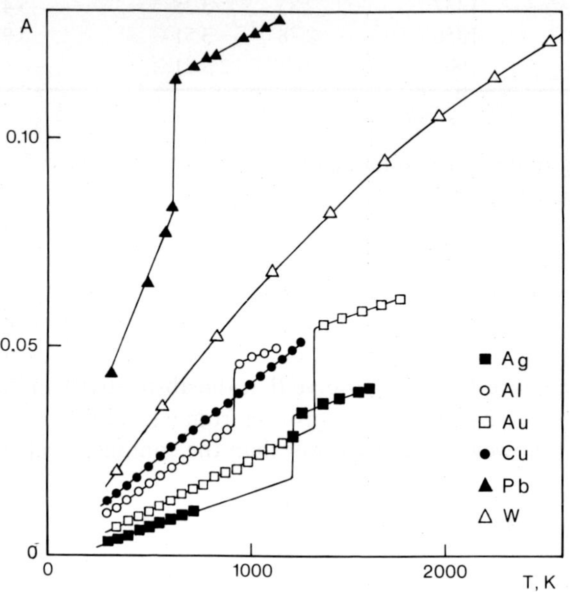

Figure 1.17 Computed curves showing the increase with temperature of the intrinsic absorptivity of some pure metals before and after melting.

1.3.3. *On the effect of 'anomalous' absorption by metals*

For some time now (see, for example, references [5, 76, 77]), it has been determined that the absorptivity of a metal submitted to pulsed power-laser irradiation exhibits, close to the melting point, a stepwise increase up to values that would significantly exceed the levels following from the previously discussed temperature dependence of the absorptivity $A(T)$, including the jump in absorptivity at the melting point. This effect is known as 'anomalous' absorption by metals.

We shall illustrate this effect with the results from papers [74, 78] where the absorptivity and reflectivity of Al was investigated on samples submitted to TEA CO_2 laser pulses.

A typical dependence of the integral (total) absorptivity per pulse, A_e, on the peak intensity I_0 in the incident laser pulse is represented in figure 1.18(*a*)

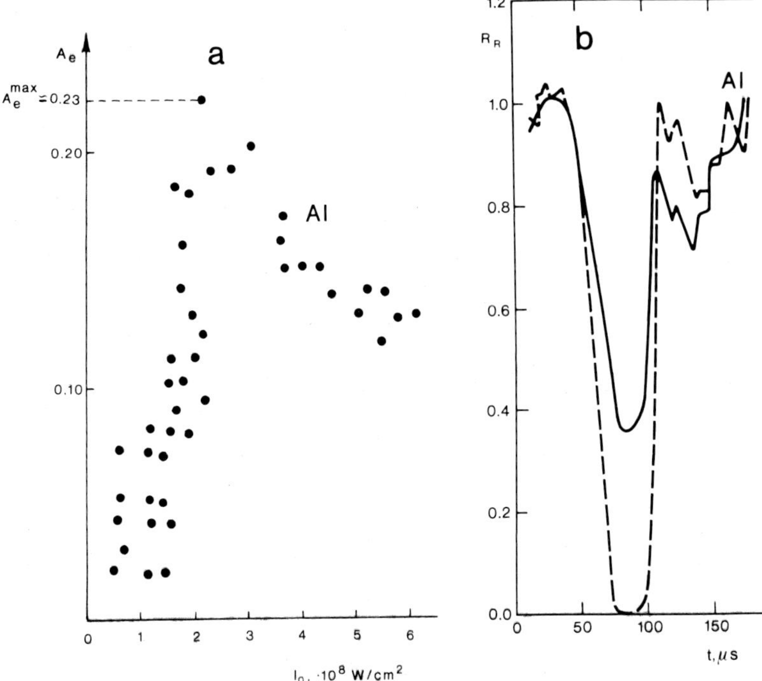

Figure 1.18 (*a*) The dependence of the integral (total) absorptivity of an aluminium sample, A_e, on peak intensity of the CO_2 laser radiation. (*b*) The time dependence of the specular reflectivity of an aluminium sample for the first (broken curve) and second subsequent (full curve) TEA CO_2 laser pulses directed on the same irradiation site, entailing sample melting in a surface layer.

(similar curves can be found in references [79–81]). We notice the jump in A_e by practically one order of magnitude within a quite narrow range of intensity values, $I_0 = (1-2) \times 10^8\ \mathrm{W\,cm^{-2}}$. Numerical evaluations, the investigation of the irradiated surface and several specially devised experiments have shown that the melting of the sample starts at $I_0 = I_m = 1.2 \times 10^8\ \mathrm{W\,cm^{-2}}$, whereas intense vaporisation of the sample surface and plasma formation begin at $I_0 = I_v = 1.7 \times 10^8\ \mathrm{W\,cm^{-2}}$. Thus the effect in this case is not determined only by heat transfer from plasma to the metal target.

The CRO traces of variation of the specular reflectivity, R_R, of an aluminium sample during the process of laser irradiation are shown in figure 1.18(*b*). The broken curve was obtained with the first laser pulse of an intensity $I_0 = 4.1 \times 10^8\ \mathrm{W\,cm^{-2}}$—corresponding precisely to the melting threshold in the experiments reported in reference [78]. We notice that the R_R value decreases to zero when surface melting begins. After irradiation no modification or other alterations of the polished surface profile were observed. The full curve corresponds to the repeated (second) irradiation of the same site on the surface, at the same fixed value of intensity, I_0. A significant variation in R_R was also observed in this case, although, in contrast with the first case, there was no plasma ignition in the irradiation zone.

The variations in A_e and R (assuming $R \simeq R_R$) close to the melting point of aluminium, as seen in figure 1.18, are too large to be accommodated in the explanations given to interpret the behaviour of aluminium absorptivity at melting.

For an appropriate interpretation of this 'anomalous' absorption one has to consider a whole series of other processes evolving during the action of intense laser irradiation of metallic surfaces. Most important among these are the dynamical surface thermal deformations, the occurrence of resonant surface periodical structures and the heat transfer to the sample, from the plasma in contact with its surface.

Chapter 2 Basic Regimes of the Heating of Metal Targets by Laser Irradiation

Convenient analytic formulae are given for calculating the evolution in space and time of the temperature field inside metal targets under various irradiation conditions and target geometries. The variation with temperature of the optical and thermophysical properties of the metal—including the transition through the melting point—is given particular emphasis.

In this chapter we intend to discuss the most typical regimes of laser heating of metals, in such conditions that no compound of any kind is formed as an effect of laser irradiation, and also no intense vaporisation or plasma generation occurs close to the surface. In other words, our reference term is the low-intensity laser irradiation of metals in inert gases.

We shall concentrate on the influence upon the laser heating dynamics of the geometrical shape and dimensions of samples, of the dimensions of the irradiation spot and the energy distribution across the laser beam, of the law governing the radiation intensity variation in time and of the beam displacement on the sample surface.

In dealing with the chief thermophysical problems we have proceeded, as a rule, from the cases discussed in reference [82], attempting an investigation of a series of problems that were not considered in the monographs [2, 3, 83], such as the influence of the actual time-shapes of the radiation pulses, the approach to the thermoconduction problem by taking into account the temperature dependence of thermophysical and optical properties, and others.

For fast numerical evaluations on the basis of the expressions given in this chapter one can use the data in the appendix, giving thermophysical constants for a series of common metals.

2.1. The laser heating source

Although the energy distribution in a cross section of the laser beam may differ from case to case, in practice one usually assumes that the following approximations can be safely assumed.

First, one takes that the energy distribution is uniform within a circular spot of radius R_s

$$I(r) = \begin{cases} I_0 & \text{for } r \leqslant R_s \\ 0 & \text{for } r > R_s \end{cases} \tag{2.1}$$

which gives a good approximation of highly multimode laser radiation.

Second comes the Gauss distribution. Typical profiles of intensity and fluence distributions, in this case corresponding to monomode laser radiation, are presented in figure 2.1. Two forms are customarily used to describe the intensity $I(r)$ or energy density $E_s(r)$ Gaussian distribution functions, differing by the characteristic radius value. When writing

$$I(r) = I_0 \exp(-r^2/(R_s^e)^2) \tag{2.2}$$

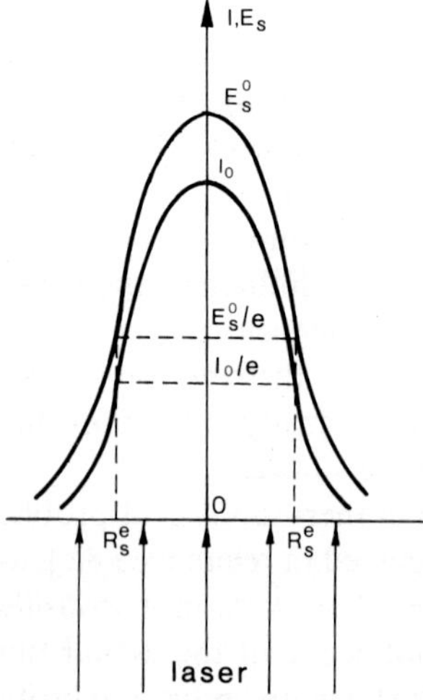

Figure 2.1 Gaussian distributions of intensity, I, and of fluence, E_s, in a cross section of a laser beam.

the characteristic dimension of the light spot, $R_s = R_s^e$ is determined by the radius at which the intensity and the fluence are e times lower than at the beam axis. Alternatively, one takes

$$I(r) = I_0 \exp(-2r^2/(R_s^{e^2})) \tag{2.3}$$

where at a radius $R_s = R_s^e$, the intensity (fluence) is e^2 times lower than at the beam axis. By comparing equations (2.2) and (2.3) one simply infers the relation $R_s^{e^2} = \sqrt{2}\, R_s^e$.

2.2. The heat conduction equation

Let us assume that the metallic sample is a homogeneous and isotropic medium. We shall consider the surface influence only through the changes in absorptivity. Volume defects and impurities shall be ignored. Then in the case of a semi-infinite sample—the most general case from which different variants for the shape of the sample are derived by imposing appropriate limitations—the heat action of the laser radiation can be described in terms of the variable temperature, $T(x, y, z, t)$ in any particular point of the sample, and at given moments in time. This quantity obeys the heat conduction equation

$$\rho c \frac{\partial T}{\partial t} = k_T \nabla^2 T + A_v I(x, y, z, t) \tag{2.4}$$

where ρ, c, k_T are the density, the specific heat and the heat conductivity of the metal, respectively, and A_v is the fraction of the radiation energy absorbed into the sample per unit time and unit volume of metal.

Equation (2.4) allows for the evaluation of the temperature field not only during the laser pulse but also after the laser is switched off, when $I(x, y, z, t > \tau_p) = 0$.

The temperature dependence of the thermophysical parameters ρ, c, k_T, and of the optical parameter, A_v, of the metal sample makes equation (2.4) non-linear, so that its analytical solution is available only in a quite limited number of cases. However, in many circumstances of practical interest one can neglect the temperature dependence of the thermophysical parameters, which allows us to obtain rather simple analytical dependences, $T(I)$. We shall begin our analysis with such simple problems, assuming, for further simplification, that the absorptivity of the sample remains constant throughout the laser irradiation process—which is of course a much more restrictive hypothesis.

2.3. Semi-infinite metal target

The skin layer depth in the metal, δ, is a few tens to a few hundreds of ångströms for the radiation of the most powerful laser sources currently available, generating in the $\lambda \simeq 0.2-10.6\ \mu m$ spectral range.

The propagation length l_{th} of the heat wave over a time span equalling the duration τ_p of the incident laser pulse can be evaluated with the following expression

$$l_{th} = (\kappa \tau_p)^{1/2} \tag{2.5}$$

where κ is the thermal diffusivity of the metal. One can consider the laser heating source as a surface source during the whole duration of the laser pulse, whenever its thickness δ is much lower than the heat-wave depth, i.e. $\delta \ll l_{th}$. Assuming, for example, $l_{th} \leqslant 10^{-6}$ cm, one obtains from equation (2.5) that this condition stands in the case of most metals—characterised by heat diffusivity values of $\kappa \sim 0.1-1\ cm^2\ s^{-1}$—for laser pulses with a duration of less than a nanosecond.

On the other hand, the semi-infinite solid approximation can be applied whenever the thickness of the metal sample, h, and the heat-wavelength, also known as thermal diffusion depth, are related by

$$h \gg l_{th}. \tag{2.6}$$

We note that if verified at room temperature, $T = T_0$, the condition (2.6) is even more closely obeyed at larger temperatures—as the thermal diffusivity of metals is generally decreasing as the temperature increases.

The calculated values [84] for l_{th} in the case of several metals, for various laser pulse durations are given in table 2.1. One notices how wide is the range of sample dimensions for which one can consider the sample as semi-infinite. For example, for a duration $\tau_p = 1$ ns the semi-infinite sample model can be used in the laser heating of foils with a thickness of only $\sim 1\ \mu m$.

Within the framework of the semi-infinite sample model one can select two limiting cases, determined by the ratio of the thermal diffusion depth, l_{th} to the radius of the irradiation spot, R_s (figure 2.2).

In the case represented in figure 2.2(a), when $R_s \gg l_{th}$, over the whole duration of the laser pulse, the one-dimensional heat wave propagates into the sample perpendicularly to the sample surface (the lateral heat loss is insignificant).

In the case represented in figure 2.2(b), when $R_s \ll l_{th}$, the laser heat source can be taken as point like, and the heat expansion bears a three-dimensional character.

We shall consider these cases separately. They are often met in practice.

2.3.1. *The plane heat wave*

If the incident laser radiation does not vary with time, $I(t \leqslant \tau_p) = I_0$, the temperature distribution inside the metal is described at different moments

Table 2.1 Thermal diffusion depth for some metals.

Metal	l_{th} (μm)			
	$\tau_p = 1$ ns	$\tau_p = 100$ ns	$\tau_p = 1$ μs	$\tau_p = 100$ μs
Al	0.28	2.84	8.98	89.8
Cr	0.14	1.46	4.62	46.1
Cu	0.30	3.06	9.68	96.8
Au	0.31	3.10	9.80	98.0
Fe	0.15	1.48	4.68	46.8
Pb	0.17	1.37	4.33	43.3
Mo	0.21	2.06	6.51	65.1
Ni	0.12	1.25	3.95	39.5
Pt	0.14	1.42	4.49	44.9
Ag	0.37	3.70	11.70	117.0
Ta	0.14	1.40	4.43	44.3
Ti	0.09	0.87	2.75	27.5
W	0.23	2.34	7.40	74.0
U	0.10	1.01	3.19	31.9
V	0.09	0.94	2.97	29.7
Zn	0.19	1.88	5.95	59.4
Zr	0.1	0.99	3.13	31.3

in time by the following solution of equation (2.4)

$$T(z, t) = (2AI_0/k_T)\sqrt{\kappa t}\ \mathrm{i\,erfc}\ [z/2(\kappa t)^{1/2}]. \tag{2.7}$$

Here the coordinate z is directed into the sample, perpendicularly to the sample surface, so that on the surface we have $z = 0$. For $z = 0$, the solution to (2.7) becomes much simpler

$$T(0, t) = (2AI_0/k_T)(\kappa t/\pi)^{1/2}. \tag{2.8}$$

Note that for very long heating durations ($t \to \infty$), according to equation (2.8) the temperature increase on the surface, $T \sim \sqrt{t}$, would be unlimited. One has, however, to note that this does not occur in practice, owing to

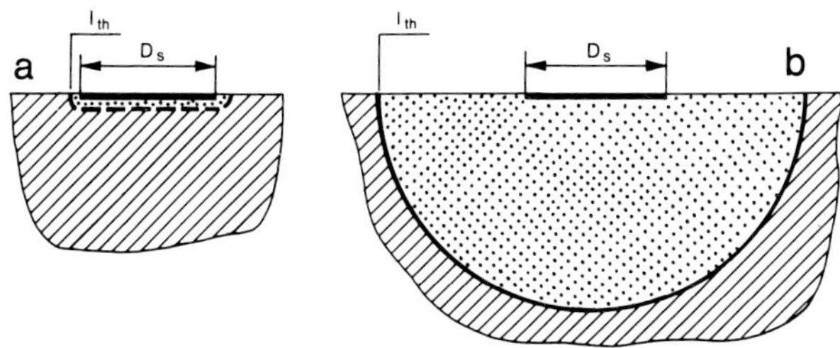

Figure 2.2 Semi-infinite samples. Approximations of a plane (a) and spherical (b) heat wave.

competition with the radiative losses from the irradiation zone and transition to three-dimensional geometry. Below the boiling point radiative losses are usually negligible in comparison with the usual mechanisms of heat conduction into the metal bulk. But such losses would rapidly increase with the fourth power of the temperature reached on the surface. That is why we have not introduced explicitly the radiative losses in equation (2.4). We shall, however, consider them later on, in more detail.

Next we shall analyse the influence of the time-shape of the laser pulse upon the dependence $T(z, t)$ (cf figure 2.3).

We have already used the rectangular shape of the radiation pulse (figure 2.3(a)) when inferring the solutions (2.7) and (2.8) of equation (2.4). After the laser pulse ceases, at $t > \tau_p$, the temperature proceeds according to the function

$$T(z, t > \tau_p) = \frac{2AI_0\kappa^{1/2}}{k_T}\left[t^{1/2}\,\mathrm{i\,erfc}\left(\frac{z}{2(\kappa t)^{1/2}}\right) - (t-\tau_p)^{1/2}\,\mathrm{i\,erfc}\left(\frac{z}{2[\kappa(t-\tau_p)]^{1/2}}\right)\right]. \tag{2.9}$$

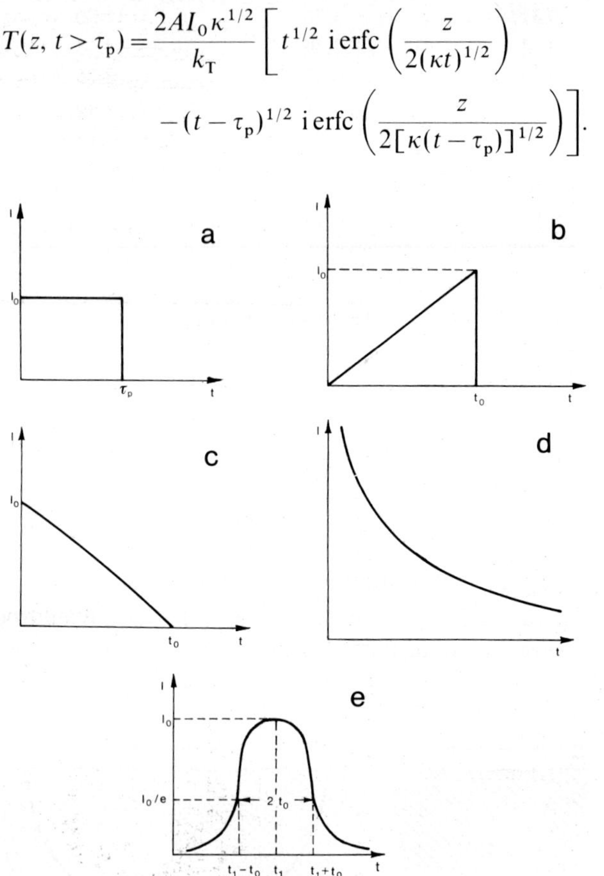

Figure 2.3 The time evolution of laser pulse intensity. (a) Rectangular pulse; (b) increasing slope; (c) decreasing slope; (d) $I(t) \sim 1/\sqrt{t}$; (e) Gaussian time pulse.

In the general case, the time evolution of temperature into the sample is described by

$$T(z, t) = \frac{A\kappa^{1/2}}{k_{\mathrm{T}}\sqrt{\pi}} \int_0^t \frac{I(t-\xi)\exp(-z^2/4\kappa\xi)}{\xi^{1/2}}\,\mathrm{d}\xi. \tag{2.10}$$

Accordingly, on the surface, $z = 0$, the temperature shows a time variation

$$T(0, t) = \frac{A\kappa^{1/2}}{k_{\mathrm{T}}\sqrt{\pi}} \int_0^t \frac{I(t-\xi)\,\mathrm{d}\xi}{\xi^{1/2}}. \tag{2.11}$$

The progress of numerical techniques now enables one to infer the temperature field inside a metallic sample for practically any time pattern of incident laser radiation intensity. As for the analytical solutions, these can be obtained only in a relatively few cases, which shall be reviewed later.

Pulses of increasing/decreasing intensity
If the intensity of the laser radiation shows a linear increase (index +) or decrease (−) in time (figures 2.3(*b*), (*c*)) described by

$$I_+(t) = I_0 t/t_0 \tag{2.12}$$

and

$$I_-(t) = I_0(1 - t/t_0) \tag{2.13}$$

where $0 \leqslant t \leqslant t_0$, it results from equation (2.11) that

$$T_+(0, t) = (4/3)\,AI_0 t^{3/2}/[t_0(\pi c\rho k_{\mathrm{T}})^{1/2}] \tag{2.14}$$

and

$$T_-(0, t) = AI_0(t_0/\pi c\rho k_{\mathrm{T}})^{1/2}(t/t_0)^{1/2}(2 - 4t/3t_0). \tag{2.15}$$

It is interesting to note [23] that both these pulses can be approximated by rectangular pulses of intensity $I(t) \simeq I_0$ and duration $\tau_{\mathrm{p}} = t_0/2$. The final temperature obtained under these circumstances is, according to equation (2.8)

$$T(0, t_0/2) = \sqrt{2}\,AI_0(t_0/\pi c\rho k_{\mathrm{T}})^{1/2}. \tag{2.16}$$

The values obtained with relation (2.16) differ by only 6% from the final temperature $T_+^{\max} = T_+(t = t_0)$ induced by a pulse linearly increasing in time—computed with the relation (2.14)—but it is 50% higher than the maximum temperature $T_-^{\max}$ on the surface of metal samples subjected to the action of a pulse of linearly decreasing intensity

$$T_-^{\max} = (2\sqrt{2}/3)\,AI_0(t_0/\pi c\rho k_{\mathrm{T}})^{1/2}. \tag{2.17}$$

In this latter case, the maximum temperature is not reached at the end of the pulse action, but at $t = t_0/2$.

The generalisation of formula (2.14) describes the time evolution of the temperature into the metal sample under the action of a linearly increasing laser pulse

$$T_{+}(z, t) = \frac{AI_0\kappa^{1/2}}{t_0 k_T}(4t)^{3/2}\mathrm{i}^3\mathrm{erfc}[z/2(\kappa t)^{1/2}] \tag{2.18}$$

where

$$\mathrm{i}^n\mathrm{erfc}\, y = \int_y^\infty \mathrm{i}^{n-1}\mathrm{erfc}\,\xi\, \mathrm{d}\xi \qquad n = 1, 2, \ldots. \tag{2.19}$$

An analytical expression for the time variation of temperature into the sample can also be established in the case when the intensity of the incident laser radiation decreases, at any $t > 0$, proportionally to $1/\sqrt{t}$ (figure 2.3(d)). Namely, one has

$$T(z, t) = \frac{AI(t)\pi^{1/2}(\kappa t)^{1/2}}{k_T}\mathrm{erfc}\left(\frac{z}{2(\kappa t)^{1/2}}\right). \tag{2.20}$$

Bell-shaped pulses

In many cases the laser pulses have a bell shape (figure 2.3(e)) which can be approximated by a Gaussian function

$$I(t) = I_0 \exp[-(t-t_1)^2/t_0^2]. \tag{2.21}$$

By operating in the relation (2.11) the change of variable, $u = \xi/t_0$, one obtains

$$T_g(0, t) = AI_0(t_0/\pi c\rho k_T)^{1/2}\int_0^{t/t_0} \mathrm{d}u^{-1/2}\exp\left[-\left(u - \frac{t-t_1}{t_0}\right)^2\right]. \tag{2.22}$$

Some values of interest obtained through numerical integration in the relation (2.22) are given in table 2.2. As one can see from this table the maximum temperature on the sample surface is reached in this case on the

Table 2.2 The adimensional temperature on the surface of a metal sample at different moments in time, when submitted to the action of a Gaussian pulse of radiation.

$(t-t_1)/t_0$	$T_g(0, t)/AI_0(t_0/\pi c\rho k_T)^{1/2}$
0 (pulse centre)	1.82
0.1	1.94
0.2	2.03
0.5	2.14
0.6 (maximum temperature)	2.15
1	1.97

decreasing part of the laser pulse while the maximum value of the integral is of the order of unity.

An approximation of the Gaussian pulse by a rectangular pulse of intensity $I(t) \simeq I_0$ and duration $\tau_p \simeq 1.67t_0$ leads to the result

$$T^{\max}(0, t = 1.67t_0) = 2.58\ AI_0(t_0/\pi c\rho k_T)^{1/2} \tag{2.23}$$

which is only 20% higher than the maximum temperature reached on the surface, under the action of a truly Gaussian pulse

$$T_g^{\max}(0, t_0) \simeq 2.15\ AI_0(t_0/\pi c\rho k_T)^{1/2}. \tag{2.24}$$

Finally we mention that metal sample heating under the action of typical pulses emitted by TEA CO_2 laser sources—consisting of a narrow peak with a halfwidth of a few hundred nanoseconds, followed by a tail that lasts for some microseconds, containing the largest part of the incident laser energy—was analysed in detail in references [85, 86].

2.3.2 The spherical heat wave

We shall introduce in this section the main relations corresponding to the case $l_{th} \gg R_s$, for different temporal–spatial patterns in which radiation is delivered.

With a time-invariable intensity and for a rectangular uniform distribution of energy inside the irradiation spot, the time variation of the temperature reached on the target surface during the laser pulse action is given by

$$T(r, z, t) = \frac{AI_0R_s}{2k_T}\int_0^\infty J_0(\xi r)J_1(\xi R_s)\left[\exp(-\xi z)\operatorname{erfc}\left(\frac{z}{2(\kappa t)^{1/2}} - \xi(\kappa t)^{1/2}\right)\right.$$
$$\left. - \exp(\xi z)\operatorname{erfc}\left(\frac{z}{2(\kappa t)^{1/2}} + (\xi t)^{1/2}\right)\right]\frac{d\xi}{\xi}. \tag{2.25}$$

Here J_0 and J_1 are the Bessel functions of the first kind, and orders zero and one, respectively.

In practice the following particular cases of equation (2.25) are noted, corresponding to the evolution of the temperature field below the centre of the irradiation spot ($r = 0$)

$$T(0, z, t) = \frac{2AI_0(\kappa t)^{1/2}}{k_T}\left[\operatorname{i\,erfc}\frac{z}{2(\kappa t)^{1/2}} - \operatorname{i\,erfc}\left(\frac{(z^2 + R_s^2)^{1/2}}{2(\kappa t)^{1/2}}\right)\right] \tag{2.26}$$

and in the centre of the irradiation spot on the sample surface ($r \simeq z \simeq 0$)

$$T(0, 0, t) = \frac{2AI_0(\kappa t)^{1/2}}{k_T}\left[\frac{1}{\sqrt{\pi}} - \operatorname{i\,erfc}\left(\frac{R_s}{2(\kappa t)^{1/2}}\right)\right]. \tag{2.27}$$

For indefinitely long laser pulses ($t \to \infty$), the maximum temperatures reached on the surface and in the sample follow from equations (2.26) and (2.27)

$$T(0, z, \infty) = \frac{AI_0}{k_T}[(z^2 + R_s^2)^{1/2} - z] \tag{2.28}$$

and

$$T(0, 0, \infty) = \frac{AI_0 R_s}{k_T} \tag{2.29}$$

while the temperature averaged over the whole area of the focal spot is related for $t \to \infty$ to the maximum temperature $T(0, 0, \infty)$ by the relation

$$\bar{T} = 0.85\, T(0, 0, \infty). \tag{2.30}$$

We note that equations (2.28)–(2.30) can be applied whenever $t \gg R_s^2/\kappa$—a condition much more easily fulfilled than the more general approximation $t \to \infty$.

Assuming that the dissipation of the incident laser energy takes place uniformly over the entire focal spot of radius R_s and occurs instantly at the time of the process

$$I(t) = I_0 \delta(t) \tag{2.31}$$

where $\delta(t)$ is a delta function of time, the time variation of temperature into the sample results from

$$T(r, z, t) = \frac{AE_0}{2c\rho\pi R_s^2(\pi\kappa^3 t^3)^{1/2}} \int_0^{R_s} \exp\left(-\frac{r^2 + r'^2 + z^2}{4\kappa t}\right) J_0^{\mathrm{m}}\left(\frac{rr'}{2\kappa t}\right) r'\, \mathrm{d}r' \tag{2.32}$$

where J_0^{m} is the modified Bessel function of zero order.

For a laser pulse of arbitrary time-shape and assuming again a rectangular profile of the energy distribution across the irradiation spot, one can use the equation

$$T(r, z, t) = \frac{A}{4c\rho\pi\kappa^{3/2}} \int_0^{R_s} \int_0^t \frac{I(r', t')}{(t - t')^{3/2}} \exp\left(-\frac{r^2 + r'^2 + z^2}{4\kappa(t - t')}\right) \times J_0^{\mathrm{m}}\left(\frac{rr'}{2\kappa(t - t')}\right) r'\, \mathrm{d}r'\, \mathrm{d}t. \tag{2.33}$$

Let us further consider Gaussian beams. Assuming again that the incident laser energy is instantly dissipated within a spot of radius R_s^e, with a Gaussian space distribution of intensity/energy density ($R_s = R_s^e$), the time evolution of temperature T_G into the sample results from an expression considerably simpler than (2.32)†

$$T_G(r, z, t) = \frac{AE_s R_s^2}{\rho c(\pi \kappa t)^{1/2}(4\kappa t + R_s^2)} \exp\left(-\frac{z^2}{4\kappa t} - \frac{r^2}{4\kappa t + R_s^2} \right). \tag{2.34}$$

Knowing the time evolution of the laser pulse, one has

$$T_G(r, z, t) = \frac{AR_s^2}{\rho c(\pi \kappa)^{1/2}} \int_0^t \frac{I_0(t')}{(t-t')^{1/2}[4\kappa(t-t') + R_s^2]} \times \exp\left(-\frac{z^2}{4\kappa(t-t')} - \frac{r^2}{4\kappa(t-t') + R_s^2} \right) \mathrm{d}t'. \tag{2.35}$$

Things are much simpler if one considers the case of incident laser intensity invariable in time. Then, in a stationary situation ($t \to \infty$), the temperature distribution inside the sample can be described by the expression

$$T_G(r, z, \infty) = \frac{AI_0 R_s^2}{2k_T} \int_0^\infty \exp(-R_s^2 \xi^2/4) \exp(-z\xi) J_0(r\xi)\, \mathrm{d}\xi. \tag{2.36}$$

In the centre of the irradiation spot, $r \simeq z \simeq 0$, we obtain the following two relations, useful in practice

$$T_G(0, 0, t) = (AI_0 R_s / k_T \sqrt{\pi}) \arctan[(4\kappa t / R_s^2)^{1/2}] \tag{2.37}$$

and

$$T_G(0, 0, \infty) = AI_0 R_s \sqrt{\pi} / 2k_T. \tag{2.38}$$

We note that formulae (2.29) and (2.38), giving the stationary temperature in the centre of the spot for a uniform and Gaussian beam, respectively, would not differ much from each other. Their ratio, in the case of identical spot radii, ($R_s = R_s^e$), is

$$T_G(0, 0, \infty)/T(0, 0, \infty) = \frac{\sqrt{\pi}}{2} \simeq 0.885. \tag{2.39}$$

If the Gaussian beam intensity $I(t)$ is of the type (2.12), i.e. an increasing slope, the time variation of the temperature in the centre of the focal spot is

† Note the difference between temperatures T_g and T_G. In the first case index g denotes a Gaussian form of the laser pulse shape, in the second case index G denotes a Gaussian form of the energy distribution in the beam.

given by

$$T_G^+(0, 0, t) = \frac{AI_0 R_s^2}{4k_T \kappa \sqrt{\pi t_0}} \{(4\kappa t/R_s^2 + 1)[\arctan(4\kappa t/R_s^2)^{1/2}] - (4\kappa t/R_s^2)^{1/2}\}. \tag{2.40}$$

Similarly, with a pulse of a linearly decreasing intensity, as described by equation (2.13), one gets

$$T_G^-(0, 0, t) = \frac{AI_0 R_s^2}{4k_T \kappa \sqrt{\pi t_0}} \{(4\kappa t_0/R_s^2 - 4\kappa t/R_s^2 - 1)[\arctan(4\kappa t/R_s^2)^{1/2}] + (4\kappa t/R_s^2)^{1/2}\}. \tag{2.41}$$

Finally let us point out that, although these relations may prove satisfactory as an introduction to the problem discussed, they cannot possibly cover entirely the variety of experimental situations. In the most difficult cases, a rapid estimation of the order of magnitude of the temperature rise within the irradiated area as an effect of laser irradiation is obtained with the expression

$$T \simeq \frac{AE_0}{\rho c V}. \tag{2.42}$$

Here E_0 is the incident laser pulse energy and ρ and c stand for the density and specific heat of the metal, respectively. Volume V submitted to the action of the heat wave until the end of the laser pulse, can be evaluated as

$$V \simeq \begin{cases} R_s^2 \sqrt{\kappa \tau_p} & \text{for } R_s \gg \sqrt{\kappa \tau_p} \\ \frac{4}{3} \pi (\kappa \tau_p)^{3/2} & \text{for } R_s \ll \sqrt{\kappa \tau_p}. \end{cases} \tag{2.43}$$

Though of a limited accuracy, equations (2.42) and (2.43) allow for the simple and fast evaluations which may be required when performing laser processing operations on metal samples.

2.4. Consideration of the temperature variation of the thermophysical and optical properties of metals

For most metals over the temperature range from room temperature to melting point, the relative variation with temperature of the thermal conductivity, k_T, thermal diffusivity, κ, and heat capacity per unit volume, C_V, does not exceed 10% [23, 84].

On the other hand, due to the very short time in which laser heating and subsequent cooling of metal samples take place, even small modifications of the thermophysical parameters—to which one has to add the much larger variation of the metal absorptivity—may entail significant changes in the temperature field obtained as a solution of the heat conduction equation.

The assumption of a temperature dependence of thermophysical and optical parameters in the heat equation (2.41) makes difficult the task of solving the problem analytically. Progress has been made, however, in the past few years in obtaining some analytical solutions for various non-linear forms of the heat equation, as well as in numerically solving these equations. Let us analyse these data following the results obtained by Uglov and co-workers [87, 88].

2.4.1 *The general (non-linear) form of the heat conduction equation*

Taking into account all the above-mentioned temperature variations, but continuing to assume that the metallic sample is homogeneous and isotropic, the general form of the heat conduction equation becomes

$$\rho(T)c(T)\frac{\partial T}{\partial t}=\frac{\partial}{\partial x}\left(k_{\mathrm{T}}(T)\frac{\partial T}{\partial x}\right)+\frac{\partial}{\partial y}\left(k_{\mathrm{T}}(T)\frac{\partial T}{\partial y}\right)+\frac{\partial}{\partial z}\left(k_{\mathrm{T}}(T)\frac{\partial T}{\partial z}\right) \tag{2.44}$$

with boundary and initial conditions

$$k_{\mathrm{T}}(T)\left.\frac{\partial T}{\partial z}\right|_{z=0}=-A(T)I(r,t) \tag{2.45}$$

$$T(\pm\infty,y,z,t)=T(x,\pm\infty,z,t)=T(x,y,\pm\infty,t)=T(x,y,z,0)=T_0. \tag{2.46}$$

Here T_0 is the initial temperature which in the above calculations was assumed to be $T_0=0$.

In cylindrical coordinates and under the assumption, natural in the context, that the temperature distribution has a cylindrical symmetry around the laser spot, equation (2.44) reads

$$\rho(T)c(T)\frac{\partial T}{\partial t}=\frac{1}{r}\frac{\partial}{\partial r}\left(k_{\mathrm{T}}(T)\frac{\partial T}{\partial r}\right)+\frac{\partial}{\partial z}\left(k_{\mathrm{T}}(T)\frac{\partial T}{\partial z}\right) \tag{2.47}$$

and conditions (2.46) become

$$T(\infty,z,t)=T(r,\infty,t)=T(r,z,0)=T_0. \tag{2.48}$$

Approximate analytical solutions of the system (2.44)–(2.46), or of the equivalent system consisting of the equations (2.45), (2.47) and (2.48) can be obtained by introducing different kinds of non-linearities.

2.4.2. *The use of average values of thermophysical parameters*

The closest case to the linear one is when one takes average values of the thermophysical parameters of the material over the considered temperature range. The problem consists in a correct choice of the average values. Indeed, the straightforward arithmetic averaging

$$\bar{k}_T = \frac{k_T(T_0) + k_T(T)}{2} \tag{2.49}$$

as well as the more elaborate calculations of the type

$$\bar{k}_T = \frac{1}{T - T_0} \int_{T_0}^{T} k_T(\xi)\, d\xi \tag{2.50}$$

do not take into consideration the marked non-linear character of metal heating by laser irradiation.

A solution would be, of course, to do the averaging starting from the actual heating curves of the metal samples under the action of laser radiation precisely in the conditions under investigation, and to use the data available from the literature regarding the variation with temperature of the thermophysical parameters of the metal. And yet a question remains open: to what extent can the parameters thus obtained be used for other, different, irradiation conditions?

Several approximations for k_T and C_V proved, however, quite universal and useful for the analysis of a large class of problems in the laser heating of metals. One possibility, for instance, is to introduce in the heat conduction equation k_T and C_V different functions of temperature, to describe the parameters of interest.

2.4.3. *Linear variation with temperature of k_T and C_V*

For an important class of metals and alloys, the dependence on temperature of the thermal conductivity, $k_T(T)$, and of the heat capacity per unit volume, $C_V(T) = c(T)\rho(T)$, can be satisfactorily approximated only by a third-order polynomial [84].

However, in some cases of interest (several varieties of steel, including the austenitic chrome–nickel, and some metals, for example molybdenum), the functions on temperature $k_T(T)$ and $C_V(T)$ are linear over a wide temperature range

$$k_T(T) = k_0 + \beta T \tag{2.51}$$

$$C_V(T) = C_V^0 + \delta_c T. \tag{2.52}$$

The linear approximations (2.51), (2.52) can also be used in the case of other metals, certainly over narrower ranges of temperature.

Under these assumptions, the problem of heating a metal submitted to the action of Gaussian light beams has been analysed in [87]. As a first step, only the linear variation with temperature of the thermal conductivity $k_T(T)$ was taken into consideration according to equation (2.51), while the variation with temperature of the heat capacity per unit volume, C_V, and of the metal sample absorptivity, A, are ignored.

Introducing the dimensionless temperature

$$T_G = \frac{AI_0}{k_0\sqrt{w_0}}u + T_0 \tag{2.53}$$

together with the quantities

$$r = \frac{1}{\sqrt{w_0}}\rho \qquad z = \frac{1}{\sqrt{w_0}}\xi \qquad h = \frac{AI_0\beta}{k_0^2\sqrt{w_0}} \qquad t = \frac{\tau}{w_0\kappa} \tag{2.54}$$

where $w_0 = (1/R_s^e)^2$, and changing the variable

$$v = u + \frac{hu^2}{2} \tag{2.55}$$

the system (2.45), (2.47), (2.48) becomes

$$\frac{1}{\rho}\frac{\partial}{\partial\rho}\left(\rho\frac{\partial v}{\partial\rho}\right) + \frac{\partial^2 v}{\partial\xi^2} = \frac{1}{(1+2hv)^{1/2}}\frac{\partial v}{\partial\tau} \tag{2.56}$$

$$\frac{\partial v}{\partial\xi} = -\exp(-\rho^2) \qquad \text{for } \xi = 0 \tag{2.57}$$

$$v(\rho, \xi, 0) = v(\infty, \xi, \tau) = v(\rho, \infty, \tau) = 0. \tag{2.58}$$

The system (2.56)–(2.58) was solved through reiterated approximations [87]. Thus, if one takes $\bar{D}$ as the average value of $D = (1+2hv)^{-1/2}$ over the temperature range considered (for example, for a typical steel blend over the temperature range 77–1875 K—between the boiling point of liquid nitrogen and the metal melting point—$\bar{D} \simeq 0.35$) one can design the following iteration process:

$$\Delta v_n = \bar{D}\frac{\partial v_n}{\partial t} + [(1+hv_{n-1})^{-1/2} - \bar{D}]\frac{\partial v_{n-1}}{\partial\tau} \tag{2.59}$$

$$\frac{\partial v_n}{\partial\xi} = -\exp(-\rho^2) \qquad \text{for } \xi = 0 \tag{2.60}$$

$$v(\rho, \xi, 0) = v(\infty, \xi, \tau) = v(\rho, \infty, \tau) = 0. \tag{2.61}$$

As a zero-order approximation one takes $v_0 = 0$. One obtains the following successive approximations

$$v_1(\rho, \xi, \tau) = (2/\sqrt{\pi})\int_0^{\sqrt{\tau/D}} \exp\left[-\left(\frac{\rho^2}{1+4x^2} + \frac{w_0}{4x^2}\right)\right]\frac{\mathrm{d}x}{1+4x^2} \tag{2.62}$$

$$v_2(\rho, \xi, \tau) = v_1(\rho, \xi, \tau) + \omega(\rho, \xi, \tau) \tag{2.63}$$

with

$$\omega(\rho, \xi, \tau) = 2\pi \int_0^\tau \int_0^\infty \int f(\rho', \xi', \tau') \left(\frac{\sqrt{D}}{2\sqrt{\pi(\tau-\tau')}} \right)^3 J_0\left(\frac{D\rho\rho'}{4(\tau-\tau')} \right)$$

$$\times \exp\left(-\frac{D(\rho^2+\rho'^2)}{4(\tau-\tau')} \right) \left[\exp\left(-\frac{D(\xi-\xi')}{4(\tau-\tau')} \right) \right.$$

$$\left. + \exp\left(-\frac{D(\rho^2+\rho'^2)}{4(\tau-\tau')} \right) \right] \rho' \, d\rho' \, d\xi' \, d\tau' \tag{2.64}$$

where

$$f = [1 - D^{-1}(1 + hv_1)^{-1/2}] \frac{v_1}{\tau}. \tag{2.65}$$

Here $J_0(u)$ is the Bessel function of the first order, zero index and imaginary argument.

In the centre of the irradiation spot these approximations give

$$v_1(0, 0, \tau) = \frac{1}{\sqrt{\pi}} \arctan \frac{2\tau^{1/2}}{(\bar{D})^{1/2}} \tag{2.66}$$

and

$$v_2(0, 0, \tau) = v_1(0, 0, \tau) + \omega(0, 0, \tau) \tag{2.67}$$

with the parameter $\omega(0, 0, \tau)$ determined from

$$\omega(0, 0, \tau) = 4\pi \int_0^\tau \int_0^\infty \int f(\rho', \xi', \tau') \left(\frac{D^{1/2}}{2\sqrt{\pi(\tau-\tau')}} \right)^3$$

$$\times \exp\left(-\frac{D(\xi'^2+\rho'^2)}{4(\tau-\tau')} \right) \rho' \, d\rho' \, d\xi' \, d\tau'. \tag{2.68}$$

In this way, the temperature—indicated by the dimensionless variable, u, —can be obtained from equation (2.55), and the heating rate can be described by the evolution of the quantity $du/d\tau$. To illustrate this, figure 2.4 shows the time evolution of the temperature variation rate in the centre of the irradiation spot, $du(0, 0, \tau)/d\tau$, on a steel sample, laser heated in the temperature range 77–1875 K, calculated starting from equation (2.66) (curve 1). For comparison, the evolution of the same quantity in the linear case was indicated, when using a value of the thermal conductivity averaged according to equation (2.47) over the temperature range examined (curve 2).

From figure 2.4, one can notice that consideration of the temperature dependence of the thermal conductivity of a metal results in a substantial modification of the rate of temperature variation—and correspondingly, of the metal sample temperature—as compared with the evolutions predicted by calculations with a constant thermal conductivity of the metal (curve 3).

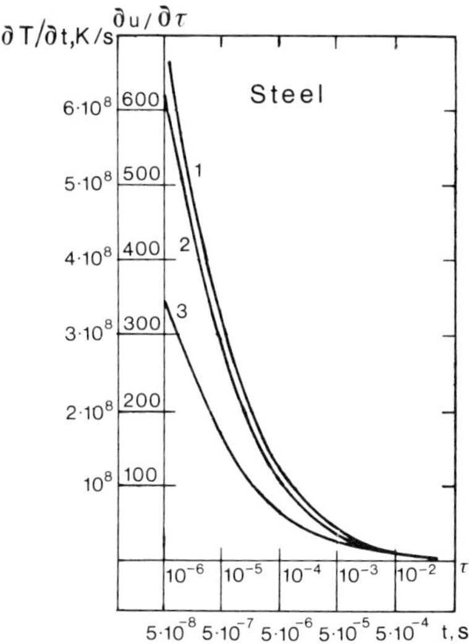

Figure 2.4 Time dependence of dimensionless laser heating rate $\mathrm{d}u(0, 0, \tau)/\mathrm{d}\tau$ in the centre of a steel sample.

In the example discussed, during the early stages of the process, $\mathrm{d}u/\mathrm{d}\tau$ values differ by a factor in excess of 2.5. However, gradually, as the duration increases (towards $\tau \geqslant 10^{-2}$), the curves 1, 2 and 3 mutually converge.

One should also mention that the approximations of first and second order for v, in solving the non-linear problem, differ in the examined case represented in figure 2.4 by less than 5%, which permits the use, in practice, of the much simpler first-order formulae (2.62) and (2.66). This situation may not be general, so that it may appear necessary to perform an appropriate numerical calculation.

Let us consider now a more complex problem [88]: together with the dependence $k_{\mathrm{T}}(T)$, we shall take also into account the function of temperature $C_{\mathrm{V}}(T)$, according to equation (2.52), while continuing to ignore the variation with temperature of the metal sample absorptivity, i.e. $A(T) = \mathrm{const}$.

The substitutions (2.53)–(2.55) are then used, with the exception of the definition for τ, which now becomes

$$\tau = \frac{w_0 k_0 t}{C_{\mathrm{V}}^0}. \tag{2.69}$$

Let us introduce parameter g defined as

$$g = \frac{A I_0 \delta_{\mathrm{c}}}{k_0 \sqrt{w_0 C_{\mathrm{V}}^0}}. \tag{2.70}$$

Then equation (2.56) is slightly modified, to read

$$\frac{1}{\rho}\frac{\partial}{\partial\rho}\left(\rho\frac{\partial v}{\partial\rho}\right)+\frac{\partial^2 v}{\partial\xi^2}=W(v, h, g)\frac{\partial v}{\partial\tau} \tag{2.71}$$

with

$$W(v, h, g)=\frac{g}{h}+\frac{1-g/h}{(1-2hv)^{1/2}} \tag{2.72}$$

whereas the next two equations (2.57) and (2.58) remain unchanged.

Following the same procedure, i.e. denoting by $\bar{W}$ the average value of the function $W(v, h, g)$ over the investigated domain (for example, in the range between room temperature and melting point $\bar{W}=0.2169$ in the case of molybdenum and 0.737 in the case of stainless steel), the following iteration process is designed

$$\Delta v_n=\bar{W}\frac{\partial v_n}{\partial\tau}+[g/h+(1-g/h)(1+2hv_{n-1})^{1/2}-\bar{W}]\frac{\partial v_{n-1}}{\partial\tau} \tag{2.73}$$

$$\frac{\partial v_n}{\partial\xi}=-\exp(-\rho^2)\qquad\text{for }\xi=0 \tag{2.74}$$

$$v_n(\rho, \xi, 0)=v_n(\infty, \xi, \tau)=v_n(\rho, \infty, \tau)=0. \tag{2.75}$$

Taking again as zero-order approximation $v_0=0$, the first-order approximation coincides with (2.62) and (2.66), respectively, where $\bar{W}$ substitutes for $\bar{D}$ (similarly the second-order approximation is obtained, etc), that is

$$v_1(\rho, \xi, \tau)=\frac{2}{\sqrt{\pi}}\int_0^{\sqrt{\tau/\bar{W}}}\exp\left[-\left(\frac{\rho^2}{1+4x^2}+\frac{w_0^2}{4x^2}\right)\right]\frac{\mathrm{d}x}{1+4x^2} \tag{2.76}$$

and

$$v_1(0, 0, \tau)=\frac{1}{\sqrt{\pi}}\arctan\frac{2}{\sqrt{W}}\sqrt{\tau}. \tag{2.77}$$

Second-order and higher approximations (which have a highly complex analytical form) can often be neglected in practice.

The results of the numerical calculations carried out with the help of data in references [89, 90] using the first-order approximation (2.77), in the cases of stainless steel and molybdenum, are given in figures 2.5 and 2.6, in the form of time dependences of the heating rate $\mathrm{d}u/\mathrm{d}\tau$ (curves 1) and of the temperature (curves 2) in the centre of the irradiation spot for three different levels of radiation intensity, I_0. For comparison, one gives the results obtained when both variations with temperature $k_T(T)$ and $C_V(T)$ are considered

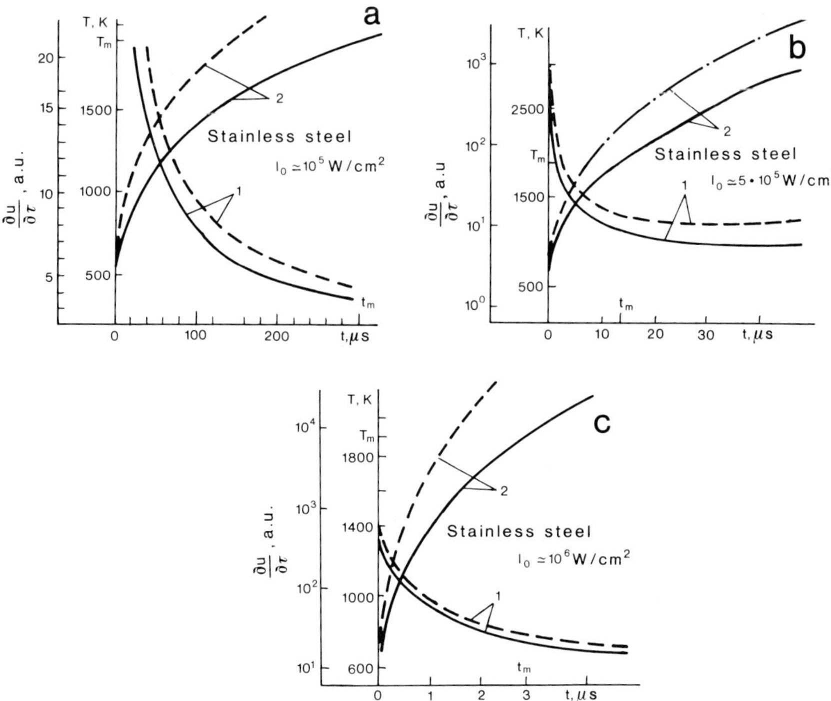

Figure 2.5 Dependence on time, t, of the dimensionless heating rate, $du/d\tau$ (curves 1) and of the temperature on the surface, T (curves 2) for stainless steel samples, for three different values of radiation intensity. Broken and chain curves correspond to the case when only the temperature dependence of the thermal conductivity $k_T(T)$ is considered, while full curves correspond to the case when the temperature dependence of the heat capacity per unit volume $C_V(T) = \rho(T)c(T)$ is also taken into account.

(full curves), and those which correspond to the case when only the dependence $k_T(T)$ is considered, while $C_V(T)$ is constant (broken curves).

It is noted that the taking into consideration of the temperature dependence of the heat capacity per unit volume, $C_V(T)$, brings about new and significant changes in the assessment of the heating kinetics of metal samples by laser irradiation, especially for the final heating stages when ignoring the $C_V(T)$ dependence leads to an overestimation (by 500 K) of the temperature reached on the sample surface.

2.4.4. *The temperature dependence of the absorptivity of metal samples*

The $A(T)$ dependence in the laser heating of metals was considered for the first time by Libenson *et al* [91]. We shall mainly base our analysis on the results of Sparks and Loh [23, 92].

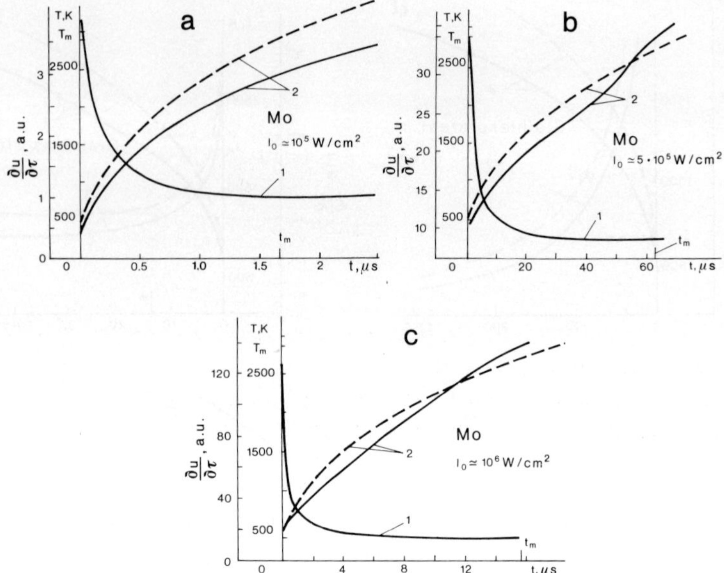

Figure 2.6 Dependence on time, t, of the dimensionless heating rate, $\mathrm{d}u/\mathrm{d}\tau$ (curves 1) and of the temperature on the surface, T, (curves 2) for molybdenum samples, for three different values of radiation intensity, when the dependence on temperature of the thermal conductivity $k_{\mathrm{T}}(T)$ and of the heat capacity per unit volume, C_{V}, are not considered (broken curves) and when their temperature dependence is taken into consideration (full curves).

Let us assume the dependence $A(T)$ to be linear, according to equation (1.95), and let the temperature dependence of the thermophysical parameters be ignored at this stage. Then, during the action of a laser pulse of rectangular time and space profile, the solution of the heat conduction equation

$$k_{\mathrm{T}} \frac{\partial^2 T}{\partial z^2} = C_{\mathrm{V}} \frac{\partial T}{\partial t} \tag{2.78}$$

with boundary and initial conditions

$$T(0, z, 0) = T_0 \qquad - k_{\mathrm{T}} \left. \frac{\partial T}{\partial z} \right|_{z=0} = A_0 I_0 + A_1 I_0 T(0, t) \tag{2.79}$$

has, in the centre of the irradiation spot ($r = 0$), the form

$$T(0, z, t) = -\frac{A_0}{A_1} + \left(T_0 + \frac{A_0}{A_1} \right) \left[\operatorname{erf} \frac{z}{2(k_{\mathrm{T}} t / C_{\mathrm{V}})^2} + \exp(-I_0 A_1 z / k_{\mathrm{T}} + u^2) \operatorname{erfc}\left(\frac{z}{2(k_{\mathrm{T}} t / C_{\mathrm{V}})} - u \right) \right] \tag{2.80}$$

where

$$u = A_1 I_0 (t/k_T C_V)^{1/2}. \tag{2.81}$$

Here

$$\text{erf } y = (2/\pi)^{1/2} \int_0^y \exp(-\xi^2)\,d\xi = 1 - \text{erfc } y. \tag{2.82}$$

On the sample surface, $z = 0$, expression (2.80) takes a simpler form, reading

$$T(0, 0, t) = (T_0 + A_0/A_1)\exp(u^2)(1 + \text{erf } u) - A_0/A_1. \tag{2.83}$$

For a constant surface absorptivity $A = A_0$, $A_1 = 0$, the expression (2.83) is then reduced to equation (2.8).

For the calculation of temperature of the irradiated surface using expression (2.83) it is important to estimate numerically the quantity

$$f(u) = \exp(u^2)(1 + \text{erf } u). \tag{2.84}$$

A semilogarithmic diagram is reproduced in figure 2.7, which can be used to estimate the function $f(u)$ by interpolation. As can be noted from the figure, the following approximations can also be applied

$$f(u) \equiv \begin{cases} 1 + 2u/\sqrt{\pi} & \text{for } u \ll 1 \\ 2\exp(u^2) & \text{for } u \gg 1. \end{cases} \tag{2.85}$$

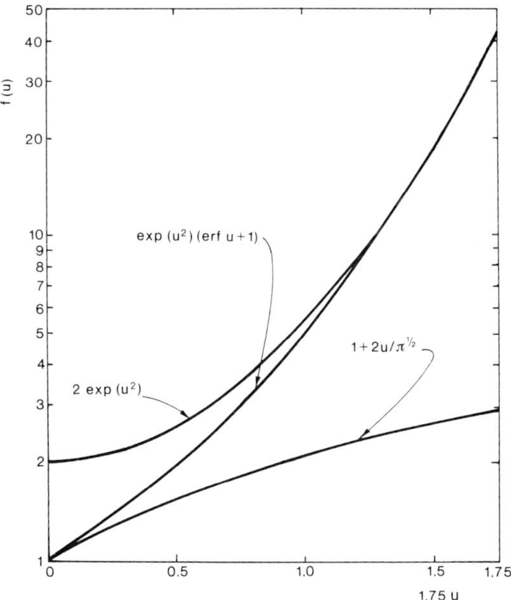

Figure 2.7 Function $f(u)$, and the approximations according to equation (2.85).

Without considering details at this point (a thorough presentation of the subject is given in Chapter 4) let us note that consideration of the variation of absorptivity with temperature results in finding surface temperature values that are significantly higher.

2.4.5. *Consideration of all time dependences—$k_T(T)$, $C_V(T)$ and $A(T)$*

Considering the time dependence of both thermophysical and optical properties of the metal, the system (2.44)–(2.46) was solved in reference [93] in the case of a Gaussian distribution of the laser intensity across the incidence spot on the irradiated surface.

The description of the evolution with temperature of the thermal conductivity, k_T, as well as of the heat capacity per unit volume, C_V, was performed by means of cubic polynomials which enable, for the majority of metals, a much better fit of their behaviour than the previously applied linear approximations (2.51) and (2.52). The evolution with temperature of the absorptivity of metal samples was described through a linear dependence of the type (1.95). Dimensionless variables were then introduced

$$u = \frac{(T - T_0)k_0\sqrt{w_0}}{I_0} \qquad x_1 = x\sqrt{w_0} \qquad y_1 = y\sqrt{w_0}$$

$$z_1 = z\sqrt{w_0} \qquad \tau = w_0\kappa_0 t. \tag{2.86}$$

Here 0 indices specify the initial values (at room temperature, T_0) of the respective physical quantities. The following transformation was also performed

$$v = \frac{A_0}{k_0}\int_0^u \frac{\kappa(x)}{A(x)}\,\mathrm{d}x. \tag{2.87}$$

As before, an iterative process was used, with the first approximation, v_1, determined from the expression

$$v_1(x_1, y_1, z_1, \tau) = \frac{A_0}{\sqrt{\pi}}\int_0^{2\sqrt{\tau/\bar{W}}} \exp\left[-\left(\frac{x_1^2 + y_1^2}{1+\xi^2} + \frac{z_1^2}{\xi^2}\right)\right]\frac{\mathrm{d}\xi}{1+\xi^2} \tag{2.88}$$

with

$$W(v) = \kappa_0/\kappa(v) \qquad \bar{W} = \frac{W_{\max} + W_{\min}}{2}. \tag{2.89}$$

It is noted that for the new variables expression (2.88) is equivalent to expression (2.35).

The second, quadratic, approximation results subsequently in the form

$$v_2(x_1, y_1, z_1, \tau) = v_1(x_1, y_1, z_1, \tau) + \omega(x_1, y_1, z_1, \tau) \tag{2.90}$$

where

$$\omega(x_1, y_1, z_1, \tau) = \frac{(\bar{W})^{1/2}}{8\pi^{3/2}} \int_{-\infty}^{\infty} \int_{-\infty}^{\infty} \int_{0}^{\infty} \int_{0}^{\tau} \frac{\psi(\alpha, \beta, \gamma, \delta)}{(\tau - \delta)^{3/2}} \times \left[\exp\left(-\frac{\bar{W}(z_1 - \gamma)^2}{4(\tau - \delta)} \right) + \exp\left(-\frac{\bar{W}(z_1 + \gamma)^2}{4(\tau - \delta)} \right) \right] \times \exp\left(-\frac{(y_1 - \beta)^2 + (x_1 - \alpha)^2}{4(\tau - \delta)/\bar{W}} \right) \mathrm{d}\alpha \, \mathrm{d}\beta \, \mathrm{d}\gamma \, \mathrm{d}\tau. \tag{2.91}$$

Here

$$\psi(x_1, y_1, z_1, \tau) = \psi_1(x_1, y_1, z_1, \tau) + \psi_2(x_1, y_1, z_1, \tau) \tag{2.92}$$

with

$$\psi_1(x_1, y_1, z_1, \tau) = -(W(v_1) - \bar{W}) \frac{\partial v_1}{\partial \tau} \tag{2.93}$$

and

$$\psi_2(x_1, y_1, z_1, \tau) = A^{-1}(v_1) \frac{\mathrm{d}A(v_1)}{\mathrm{d}v_1} \left[\left(\frac{\partial v_1}{\partial x_1} \right)^2 + \left(\frac{\partial v_1}{\partial y_1} \right)^2 + \left(\frac{\partial v_1}{\partial z_1} \right)^2 \right]. \tag{2.94}$$

The most important contribution to the value of the quantity $\psi(x_1, y_1, z_1, \tau)$ is made by the terms

$$(W(v_1) - \bar{W}) \frac{\partial v_1}{\partial \tau} \qquad \text{and} \qquad A^{-1}(v_1) \frac{\mathrm{d}A(v_1)}{\mathrm{d}v_1} \left(\frac{\partial v_1}{\partial z_1} \right)^2.$$

Then from (2.91), using equations (2.92)–(2.94), one obtains

$$\omega_1(x_1, y_1, z_1, \tau) = \frac{A_0 \sqrt{\tau}}{\sqrt{\pi \bar{W}}(1 + 4\tau/\bar{W})} \left(1 - \frac{1}{2\bar{W}} - \frac{W(v_1(0, 0, 0, \tau))}{2\bar{W}} \right) \times \exp\left(-\frac{x_1^2 + y_1^2}{1 + 4\tau/\bar{W}} - \frac{z_1^2}{4\tau/\bar{W}} \right). \tag{2.95}$$

The analytical calculation of the quantity ω_2 which, similarly, corresponds to the function ψ_2, was possible only on the assumption of certain severely simplifying hypotheses, the most limiting of these being $z = 0$ and $t \ll (w_0 x_0)^{-1}$. The use of the approximation

$$v_2(x_1, y_1, z_1, \tau) \simeq v_1(x_1, y_1, z_1, \tau) + \omega_1(x_1, y_1, z_1, \tau) \tag{2.96}$$

enables one to estimate the temperature field in a metal subjected to laser action with a higher accuracy than the solution offered by any linear model.

Formula (2.90) can give a remarkably simple expression in the centre of the irradiation spot, with the assumption $\tau \ll 1$. In this case the following relation—easy to handle in numerical computations of practical interest—is obtained

$$\begin{aligned} v_2(0,0,0,\tau) &= v_1(0,0,0,\tau)+\omega_1(0,0,0,\tau)+\omega_2(0,0,0,\tau) \\ &\simeq \frac{A_0}{\sqrt{\pi}}\left(\tau_1^{1/2}-\frac{\tau_1^{3/2}}{3}+\frac{\tau_1^{5/2}}{5}\right) \\ &\quad +\frac{A_0}{\sqrt{\pi \bar{W}}}\left(1-\frac{1}{2\bar{W}}-\frac{W(v_1(0,0,0,\tau))}{2\bar{W}}\right) \\ &\quad \times \frac{\sqrt{\tau}}{1+\tau_1}+\frac{\mu A_0 I_0}{4\pi k_T\sqrt{w_0}}\left[\pi-2+\left(\frac{4}{9}-\frac{\pi}{12}\right)\tau_1^2\right]\tau_1 \end{aligned} \tag{2.97}$$

where

$$\tau_1 = 4\tau/\bar{W} \qquad \bar{k}_T = (k_T^{max}+k_T^{min})/2 \tag{2.98}$$

$$\mu = dA(T)/dT. \tag{2.99}$$

Results of calculations, performed by means of formulae (2.97)–(2.99), of the temperature evolution at the centre of a Gaussian irradiation spot on the surface of tungsten and molybdenum samples are presented in figure 2.8. Together with the results obtained with the above mentioned algorithm (curves 1), the results of numerical computations based on the system (2.44)–(2.46) (curves 2) are presented, as well as the results of the linear theory based on either average values (curves 3), or room temperature (curves 4), of the thermophysical constants and absorptivity of metal samples.

First, note that the analytical solution above is in good agreement with numerical calculations of the non-linear problem.

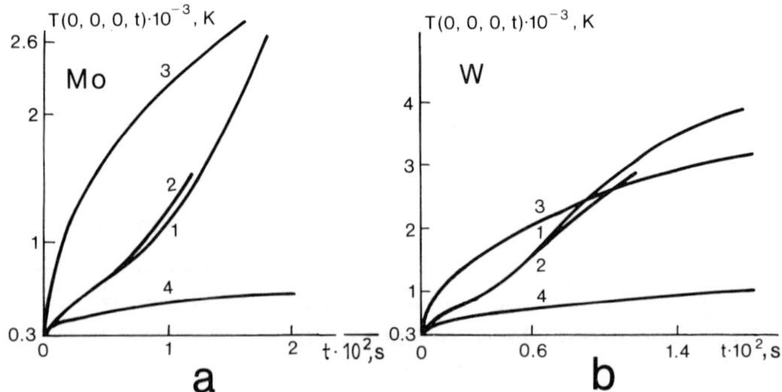

Figure 2.8 Time evolution of temperature in the centre of the irradiation spot on the surface obtained using different computational methods for molybdenum (*a*) and tungsten (*b*) samples; $I_0 = 2\times10^5$ W cm^{-2}.

Secondly, one sees that the solutions of the non-linear system agree with those of the linear system for only a small time period, at the onset of heating.

Thirdly, when the average values of the thermophysical parameters and absorptivity within the selected temperature range are used in calculations (curves 3), the solution of the linear system exhibits much higher values as compared with the non-linear system. However, during the late stages the situation reverses, i.e. for large irradiation duration, curves 3 lie lower than curves 1 and 2. The widening of the temperature range over which the average values are obtained leads to a reduction in the amplitude of the curves 3.

We also note that, based on solutions of the linear problem, the temperature in the centre of the irradiation spot is described by a time dependence whose graphical representation does not show any inflexion points, while the solutions of the non-linear problem generally show inflexion points (curves 2, 3 on figure 2.8).

Another peculiarity consists in the change of temperature distribution profile on the surface, when passing from the solution of the linear problem to the solution of the complete, non-linear problem, i.e. when including all the time dependences. For example, the evolution of the temperature normalised against its value at the point $r = 0$, in the case of laser heating of molybdenum, is represented in figure 2.9. Curves 1 and 2 correspond, respectively, to the solutions of the non-linear and linear problems at the moment $t = 5 \times 10^{-2}$ s.

It can be noted that taking into consideration all the non-linearities results in the narrowing of the temperature distribution on the sample surface—an effect which is connected to the increase in temperature and absorptivity, simultaneously with the decrease in the thermal conductivity of the metal.

Emphasising again the good results that have been obtained, using various approximations, in solving analytically the non-linear heat conduction

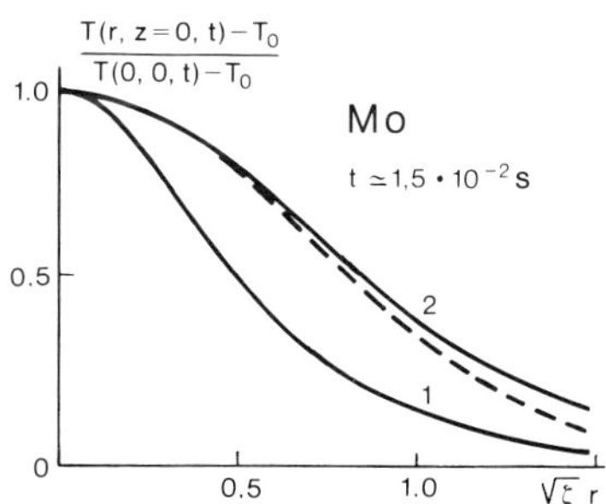

Figure 2.9 Temperature distribution in the irradiation zone on the surface of a molybdenum sample subjected to laser irradiation, $I_0 = 2 \times 10^5$ W cm^{-2}, after a time $t = 15\ \mu$s from the onset of the irradiation. Curves 1 and 2 correspond to the solutions of non-linear and linear problems, respectively; the broken curve indicates the Gaussian distribution of intensity in the incident laser beam.

equation, one must, however, admit that currently this problem is better solved numerically. There now exist comprehensive and relatively fast computer codes, which, based on the non-linear equation of heat diffusion (2.42)–(2.44), enable one to calculate with a high accuracy the temperature field in metal samples, both during laser heating and after laser irradiation. For example, a finite-difference algorithm was proposed [94], which allows one to include in the calculation, as well as the aforementioned temperature dependences, further details such as the time-shape of the laser pulse, the non-linear temperature dependence of absorptivity, etc.

2.5. Laser source in motion

A situation often encountered in practice is that of the displacement of the laser spot on the sample surface, for example when performing certain processing operations (welding, cutting, etc).

Rosenthal [95] has introduced a simple solution of the heat conduction equation which ensures a good approximation of the temperature field in a stationary regime

$$T(r, x) = \frac{AP}{2\pi k_T r} \exp\left(-\frac{vr}{2\kappa} - \frac{vx}{2\kappa}\right). \tag{2.100}$$

Here it is assumed that the laser heat source moves with a velocity v in the direction of the $0x$ axis, P is the incident power and $r = (x^2 + y^2 + z^2)^{1/2}$.

Linear approximations of the solution of the heat diffusion equation in this case can also be found in a number of works (as for example [2, 3, 96]). We shall indicate here a way of treating the most general non-linear situation, i.e. the consideration of the temperature dependences of both the thermophysical parameters and the absorptivity of metal samples [93].

In the system (2.44)–(2.46) only the first equation is modified to become

$$\frac{\partial}{\partial x}\left(k_T(T)\frac{\partial T}{\partial x}\right) + \frac{\partial}{\partial y}\left(k_T(T)\frac{\partial T}{\partial y}\right) + \frac{\partial}{\partial z}\left(k_T(T)\frac{\partial T}{\partial z}\right) = C_V(T)\left(\frac{\partial T}{\partial t} - v\frac{\partial T}{\partial x}\right). \tag{2.101}$$

The system was also solved through an iterative process, after performing the transformations (2.86) and (2.87). For example, in a first approximation one obtains

$$v_1(x_1, y_1, z_1, \tau) = \frac{A_0}{\sqrt{\pi}} \int_0^{2\sqrt{\tau/W}} \exp\left[-\left(\frac{y^2 + (x + \alpha\xi/4)^2}{1+\xi^2} + \frac{z^2}{\xi^2}\right)\right] \frac{d\xi}{1+\xi^2}. \tag{2.102}$$

The meaning of the parameters τ and $\bar{W}$ is that introduced through the relations (2.86) and (2.87), and the parameter α is determined as

$$\alpha = \frac{v\bar{W}}{\kappa_0\sqrt{W_0}}. \tag{2.103}$$

The second-order approximation can subsequently be established from the relations (2.90)–(2.92), where the expression (2.92) of the function $\psi(x_1, y_1, z_1, \tau)$ is modified so as to include a new term

$$\psi(x_1, y_1, z_1, \tau) = \psi_1(x_1, y_1, z_1, \tau) + \psi_2(x_1, y_1, z_1, \tau) + \psi_3(x_1, y_1, z_1, \tau) \tag{2.104}$$

with

$$\psi_3(x_1, y_1, z_1, \tau) = -\alpha\left(1 - \frac{W(v_1)}{\bar{W}}\right)\frac{\partial v_1}{\partial x}. \tag{2.105}$$

The analytical calculation of the corresponding correction, ω_3, was not possible without again making the approximations $z = 0$ and $t \ll (w_0 x_0)^{-1}$. It is, however, noted that in this situation also, the main contribution in estimating the second-order approximation is brought by the term ω_1, and to a lesser extent by the term ω_2. This is why we can write, in a good approximation

$$v_2(x_1, y_1, z_1, \tau) \simeq v_1(x_1, y_1, z_1, \tau) + \omega_1(x_1, y_1, z_1, \tau) \tag{2.106}$$

where the linear approximation $v_1(x_1, y_1, z_1, \tau)$ is estimated by means of equation (2.102).

2.6. Metal foils (samples of finite thickness)

In a series of cases of practical interest, for example when heating metal foils under the action of CW laser radiation, the heat wave can reach the rear (opposite to the exposed) surface during irradiation, and the model of the semi-infinite sample can no longer be used. In such circumstances, characterised by the conditions $l_{th} \simeq (\kappa\tau_p)^{1/2} \geqslant h$ (where h is the thickness of the sample) one needs equations that would differ from those recommended above, for the calculation of the temperature field in the samples. In the following we shall refer to such samples as to 'thermally thin blades (foils)'.†

Two limiting situations are encountered when dealing with the approximation of thermally thin blades: large irradiation spots, when $D_s \gg l_{th}$ (figure 2.10), and small irradiation spots, when $D_s \ll l_{th}$ (figure 2.11). We shall

† We mention that, in practicc, for moments in time during the irradiation $t \ll h^2/\kappa$ and $t \ll \tau_p$, one can further make use of the approximation of a semi-infinite sample, even when the condition $l_{th} = (\kappa\tau_p)^{1/2} \gg h$ is fulfilled.

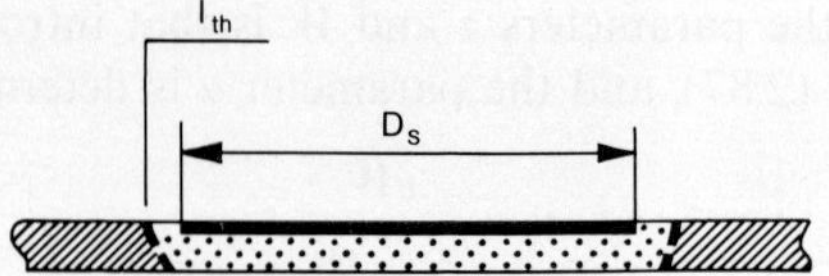

Figure 2.10 Thermally thin blade. The case when the large irradiation spot approximation ($D_s \gg l_{th}$) is to be invoked.

examine separately these two limiting situations, and also the case of forced convective cooling of metal samples.

2.6.1. *Plane heat wave*

When the condition $D_s \gg l_{th}$ is fulfilled (figure 2.10), the temperature field is uniform along the r coordinate while the temperature depends only on the z coordinate and time t.

Then, for a uniform distribution of energy within the irradiation spot, and assuming that for the surface exposed to laser irradiation we have $z = h$, the temperature inside the sample is given by the relation

$$T(z,t) = \frac{AI_0 t}{c\rho h} + \frac{AI_0 t}{k_T}\left[\frac{3z^2 - h^2}{6h^2} - \frac{2}{\pi^2}\sum_{n=1}^{\infty}\frac{(-1)^n}{n^2}\exp\left(-\frac{\kappa^2 n^2 \pi^2 t^2}{h^2}\right)\cos\frac{n\pi z}{h}\right]. \tag{2.107}$$

The first term in this expression is linear in the time t, while the second adds the effects of the multiple reflections of the heat wave.

The analytical calculations for a pulse of an arbitrary time-shape are very intricate [2, 3, 71, 82]. An important exception is the time evolution of the temperature field induced under the action of a pulse of the shape

$$I(t) = Bt^{m/2} \qquad m = -1, 0, 1, \ldots \tag{2.108}$$

namely

$$T(z, t) = \frac{2^{m+1} AI(t)\sqrt{\kappa t}\,\Gamma(m/2+1)}{k_T} \times \sum_{n=0}^{\infty}\left(\mathrm{i}^{m+1}\,\mathrm{erfc}\,\frac{(2n+1)h - z}{2\sqrt{\kappa t}} + \mathrm{i}^{m+1}\,\mathrm{erfc}\,\frac{(2n+1)h + z}{2\sqrt{\kappa t}}\right). \tag{2.109}$$

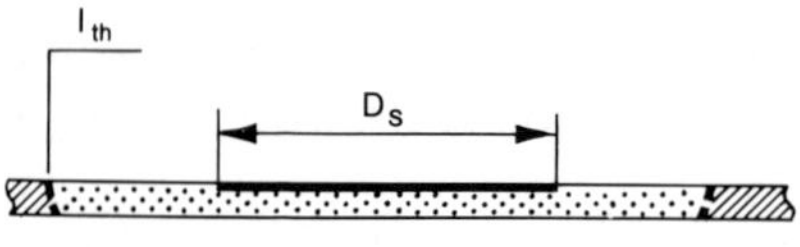

Figure 2.11 Thermally thin blade. The small irradiation spot approximation ($D_s \ll l_{th}$).

The special Γ function is defined as

$$\Gamma(z)=\int_0^\infty t^{z-1}\exp(-t)\,\mathrm{d}t. \tag{2.110}$$

In the case of thermally thin blades or discs, and particularly when these exhibit extensive surfaces, the heat losses, either radiative or convective, may become important (cf, for example, [1–3, 82, 83, 97]). Then for each surface the losses can be introduced by the expression

$$I_1(T)=\sigma_{SB}\sigma_0(T^4-T_0^4)+\eta(T-T_0). \tag{2.111}$$

Here σ_0 is the emissivity of the target surface, also called the blackening degree, η is the constant of heat convective exchange, while $\sigma_{SB}=5.75\times10^{-8}\ \mathrm{W\,m^{-2}\,K^{-4}}$ is the Stefan–Boltzmann constant. For long-enough irradiation durations, the temperature difference between the exposed surface and the opposite surface of the sample can be small, and the one-dimensional heat diffusion equation with boundary and initial conditions can be written as

$$c\rho\frac{\partial T}{\partial t}=k_{\mathrm{T}}\frac{\partial^2 T}{\partial z^2}+\alpha A I_0 \mathrm{e}^{-\alpha z} \tag{2.112}$$

$$-k_{\mathrm{T}}\left.\frac{\partial T}{\partial z}\right|_{z=0}=k_{\mathrm{T}}\left.\frac{\partial T}{\partial z}\right|_{z=h}=-I_1(T) \tag{2.113}$$

$$T(0)=T_0 \tag{2.114}$$

where α is the absorption coefficient of the target material.

By integration over the thickness of the metallic sample, $h \gg 1/\alpha$, one obtains

$$c\rho h\frac{T}{t}=-2I_1(T)+AI_0. \tag{2.115}$$

When a stationary temperature $T_s=T(t\to\infty)$ (steady-state regime) is reached, equation (2.115) becomes

$$-2I_1(T_s)+AI_0=0 \tag{2.116}$$

with the solution

$$T_s=\tfrac{1}{2}[(2\eta/\sigma_0\sigma_{SB}\sqrt{X}-X)^{1/2}-\sqrt{X}] \tag{2.117}$$

where

$$X=\left\{\frac{1}{2}\left(\frac{\eta}{\sigma_0\sigma_{SB}}\right)^2+\left[\frac{1}{4}\left(\frac{\eta}{\sigma_0\sigma_{SB}}\right)^4+\left(\frac{4a}{3}\right)^3\right]^{1/2}\right\}^{1/3}$$
$$+\left\{\frac{1}{2}\left(\frac{\eta}{\sigma_0\sigma_{SB}}\right)^2-\left[\frac{1}{4}\left(\frac{\eta}{\sigma_0\sigma_{SB}}\right)^4+\left(\frac{4a}{3}\right)^3\right]^{1/2}\right\}^{1/3}. \tag{2.118}$$

Here the parameter a is to be determined from

$$a = T_0^4 + \eta\frac{T_0}{\sigma_0\sigma_{SB}} + \frac{AI_0}{2\sigma_0\sigma_{SB}}. \tag{2.119}$$

2.6.2. Heat source of finite dimensions

The radiative energy losses become important whenever the diameter of the irradiation spot is smaller than the characteristic dimension of the heat wave (figure 2.11).

In this case, and for a uniform distribution of energy inside the irradiation spot, the temperature is to be determined, for $r = 0$, from the expression [98]

$$T(0, z, t) = \frac{AI_0(4\kappa t)^{1/2}}{k_T}\sum_{n=-\infty}^{\infty}\left(\operatorname{i\,erfc}\frac{|z-2nh|}{(4\kappa t)^{1/2}} - \operatorname{i\,erfc}\frac{[(z-2nh)^2 + R_s^2]^{1/2}}{(4\kappa t)^{1/2}}\right). \tag{2.120}$$

Analytical expressions of the temperature, useful in practice, can be obtained in the case $(\kappa\tau_p)^{1/2} \gg h$ by assuming the heat to be uniformly dissipated in a cylinder of diameter D_s and height h. Then the average temperature inside the cylinder can be evaluated from

$$T(r \leqslant R_s, z, t) = \begin{cases} \dfrac{AI_0R_s^2}{4k_Th}\left[\ln\left(\dfrac{4\kappa t}{GR_s^2}\right) + \dfrac{R_s^2}{4\kappa t}\ln\left(\dfrac{4\kappa t}{GR_s^2}\right) + \dfrac{R_s^2}{2\kappa t}\right] & \text{for } R_s^2 \ll \kappa t \\ \dfrac{AI_0R_s}{k_Th}(\kappa t)^{1/2}\left(\dfrac{1}{\sqrt{\pi}} - \dfrac{(\kappa t)^{1/2}}{4R_s}\right) & \text{for } R_s^2 \gg \kappa t \end{cases} \tag{2.121}$$

while outside the cylinder, the temperature results from the expression

$$T(r > R_s, z, t) = -\frac{AI_0R_s}{\pi k_Th}\int_0^\infty [1 - \exp(-\kappa^2 ut)] \times \left(\frac{J_0(ur)Y_1(uR_s) - Y_0(uR_s)J_1(uR_s)}{u^2(J_1^2(uR_s) + Y_1^2(uR_s))}\right)du \tag{2.122}$$

where J_0, J_1, Y_0 and Y_1 are Bessel functions of the first and second kind, and of zero and first order, respectively. Equation (2.122) can be further transformed to read

$$T(r > R_s, z, t) = \begin{cases} \dfrac{AI_0R_s^2}{4k_T}\left[\ln\left(\dfrac{4\kappa t}{Gr^2}\right) + \dfrac{R_s^2}{4\kappa t}\ln\left(\dfrac{4\kappa t}{Gr^2}\right) + \dfrac{1}{4\kappa t}\left(R_s^2 + r^2 + 2R_s^2\ln\dfrac{R_s}{r}\right)\right] \\ \text{for } R_s^2 \ll \kappa t \\ \dfrac{AI_0R_s^{3/2}}{k_Th}\left(\dfrac{\kappa t}{r}\right)^{1/2}\left[\operatorname{i\,erfc}\left(\dfrac{r-R_s}{2(\kappa t)^{1/2}}\right) - \dfrac{(3r+R_s)(\kappa t)^{1/2}}{4R_sr}\operatorname{i^2\,erfc}\left(\dfrac{r-R_s}{2(\kappa t)^{1/2}}\right)\right] \\ \text{for } R_s^2 \gg \kappa t \end{cases} \tag{2.123}$$

with

$$i^2\operatorname{erfc} y = \tfrac{1}{4}(\operatorname{erfc} y - 2y\,\mathrm{i}\operatorname{erfc} y) \tag{2.124}$$

and $G \simeq 1.781$.

Regarding the convective losses a suggestion was made [99] to substitute $I_0 - I_c$ for I_0 in equations (2.120)–(2.123). Here I_c is an intensity introduced such as to compensate for the convective losses, which can be established either empirically or by calculation, according to the given geometry and to the temperature difference between sample and environment.

Assuming that an amount of energy E_0 is instantaneously dissipated in a Gaussian spot of radius $R_s = R_s^{e^2}$, the following evolution of the temperature field on the sample surface is revealed [2, 3, 83, 100]:

$$T_G(r, t) = \left(\frac{2\pi A E_0 \exp(-\kappa k_c^2 t)\exp(-r^2/4\kappa t)}{4\pi k_T h t}\right) \times \left(\frac{1}{R_s^{-2} + (4\kappa t)^{-1}}\right)\exp\left(\frac{(r^2/16\kappa^2 t^2)^{-1}}{R_s^2 + (4\kappa t)^{-1}}\right). \tag{2.125}$$

Here the convective losses are taken into account through the term

$$k_c^2 = 2\eta/(k_T h) \tag{2.126}$$

where η is the convective losses constant (in air, $\eta \simeq 5 \times 10^{-4}\ \mathrm{W\,cm^{-2}\,K^{-1}}$). The temperature of the exposed surface and that of the opposite surface of the foil are assumed to be equal.

For a Gaussian source invariable in time, the temperature field by the end of the laser pulse action results from ($R_s \simeq R_s^{e^2}$)

$$T_G(r, \tau_p) = \frac{A I_0 \kappa R_s^2}{k_T h}\int_0^{\tau_p} \frac{dt'}{4\kappa t' + R_s^2}\exp\left(-\kappa k_c^2 t' - \frac{r^2}{4\kappa t'} + \frac{r^2 R_s^2}{4\kappa t'(R_s^2 + 4\kappa t')}\right). \tag{2.127}$$

Useful indications and tables for the numerical estimation of the integral (2.127) are given in references [2, 100].

The temperature evolution in the centre of the irradiation spot on the target is expressed through ($R_s = R_s^e$) [101]

$$T_G(0, t) = \frac{\sqrt{\pi} A I_0 R_s}{4k_T}\left\{1 + \frac{2R_s}{\sqrt{\pi} h}\left[\ln\left(\frac{k_T t}{c\rho h^2}\right) + 0.6\right]\right\}. \tag{2.128}$$

We mention that for long-enough irradiation durations

$$t \geqslant \frac{5(R_s)^2}{2\kappa} \qquad R_s = R_s^e \tag{2.129}$$

the calculation of the temperature field that results at the irradiation of a foil with a Gaussian laser beam can be performed by using equations (2.120) and (2.122) instead of equation (2.127) by taking

$$R_s = \frac{R_s^e}{1.33}. \tag{2.130}$$

2.6.3. *Forced cooling of metal samples (films)*

In many instances thin metal films are used as optical elements for lasers, e.g. mirrors, deflectors, etc. When such films are exposed to the action of power-laser beams, it might be necessary to ensure a fast removal of heat dissipated into samples following the absorption of laser radiation. Only in this way can one preserve the geometry of the exposed optical surface, while also avoiding its permanent damage, e.g. by melting.

Air, water, other gases and liquids can be used as heat carrying agents. These must be rapidly circulated in close contact with the unexposed surface of the sample. Under the hypothesis of a uniform approximation ($D_s \gg l_{th}$, $D_s \gg h$), and by assuming that the flux of absorbed laser radiation and the forced cooling flux are equal one can obtain a stationary temperature value from the expression given in reference [101]

$$T_{t\to\infty} = AI_0(l/\eta + h/k_T). \tag{2.131}$$

Here η is the constant of the convective heat exchange.

2.6.4. *Consideration of the temperature dependence of the thermophysical parameters*

The dependences on temperature of the thermal conductivity, $k_T(T)$, and of the heat capacity per unit volume, $C_V(T)$, were accounted for by means of an iterative process in [93]. Using the dimensionless quantities introduced by equations (2.53), (2.54), (2.69) and (2.70), then performing the transformation (2.55) one gives the equation system (2.45), (2.47), (2.48) the following expression:

$$\frac{1}{\rho}\frac{\partial}{\partial\rho}\left(\rho\frac{\partial v}{\partial\rho}\right)+\frac{\partial^2 v}{\partial\xi^2} = W(r, h, g)\frac{\partial v}{\partial\tau} \tag{2.132}$$

$$\frac{\partial v}{\partial\xi} = -\exp(-\rho^2) \qquad \text{for } \xi = 0 \tag{2.133}$$

$$v(\rho, \xi, 0) = v(\infty, \xi, \tau) = \frac{\partial v}{\partial\xi}(\rho, \xi = h\sqrt{k_T}, \tau) = 0. \tag{2.134}$$

By an iteration process, after several integral transformations one finds the first approximation of the solution of the system (2.132)–(2.134) in the form

$$v_1(\rho, \xi, \tau) = \frac{1}{2hk_T^{1/2}}\int_0^\infty \exp\left(-\frac{x^2}{4}\right)J_0(x\rho)\left[\!\left[\frac{[1-\exp(-x^2\bar{W}^{-1}\tau)]}{x}\right.\right. + 2\sum_{n=1}^{\infty} x\left\{1-\exp\left[-\left(x^2+\frac{n^2\pi^2}{h^2k_T}\right)\tau\bar{W}^{-1}\right]\right\}\cos\left(\frac{\pi n w_0}{hk_T^{1/2}}\right) \times\left.\left.\left(x^2+\frac{n^2\pi^2}{h^2k_T}\right)^{-1}\right]\!\right]\mathrm{d}x. \tag{2.135}$$

Here $J_0(x\rho)$ is a Bessel function of the first kind and zero order. It is shown that in evaluating the sum in the right-hand side of the expression (2.135), it is sufficient to retain only two to three terms.

For short times τ and thin foils, the relation (2.135) is simplified to read

$$v_1(\rho, \xi, \tau) \simeq hk_T^{1/2} \exp(-\rho)\left[\frac{\tau}{h^2 k_T \bar{W}} + \frac{(hk^{1/2} - w_0)^2}{2h^2 k_T} - \frac{1}{6} - 2\sum_{n=1}^{\infty} \frac{(-1)^n (1 - w_0 hk_T^2)\cos n\pi}{n^2\pi^2} \exp\left(-\frac{n^2\pi^2\tau}{\bar{W}h^2 k_T}\right)\right]. \quad (2.136)$$

By using equations (2.135) and (2.136) one can further obtain both the temperature field and the heating rate of the sample. When the modification of k_T and C_V during the heating process is taken into account starting from this approximation, the notations and transformations used in the semi-infinite sample model must be employed.

2.7. Metal layer on metal base

Many metallic samples used in experiments concerning the interaction of laser radiation with solid surfaces, as well as optical components (especially mirrors) used in conjunction with lasers are manufactured in the form of 'metallic sandwiches'. In a metallic sandwich, metal layers are deposited on a metal base plate (figure 2.12).

In this case, for a plane geometry of irradiation, and also assuming that the thickness of the skin layer is much less than the thickness h of the deposited layer, the temperature distribution into the layer along the coordinate $z \leqslant h$, measured from the surface exposed to irradiation, can be determined with the aid of the expression

$$T_1(z, t) = \frac{AI_0}{k_{T_1}}\left[(4\kappa_1 t)^{1/2} \sum_{n=-\infty}^{\infty} \xi^{|n|} \,\mathrm{i\,erfc}\left(\frac{|z - 2nh|}{(4\kappa_1 t)^{1/2}}\right)\right] \quad (2.137)$$

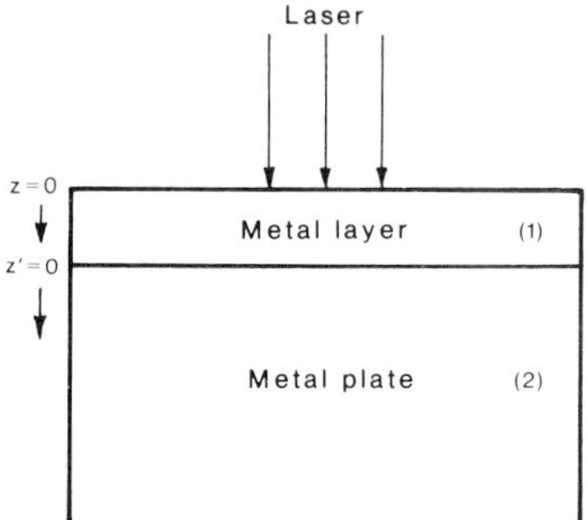

Figure 2.12 Two-layer metallic structure.

with

$$\xi = (k_{T_1}\sqrt{\kappa_2} - k_{T_2}\sqrt{\kappa_1})/(k_{T_1}\sqrt{\kappa_2} + k_{T_2}\sqrt{\kappa_1}). \tag{2.138}$$

Here k_{T_1}, k_{T_2}, κ_1 and κ_2 are the thermal conductivity and thermal diffusivity of the deposition and metallic support, respectively.

The temperature evolution inside the substrate along the coordinate z' measured from the interface results from

$$T_2(z', t) = [2T_c/(\Lambda + 1)] \sum_{n=0}^{\infty} \xi^n I_1(z^*) \tag{2.139}$$

where

$$T_c = \frac{2AI_0(\kappa_2 t)^{1/2}}{k_{T_2}\sqrt{\pi}} \qquad \Lambda = \frac{k_{T_1}}{k_{T_2}}\left(\frac{\kappa_2}{\kappa_1}\right)^{1/2} \tag{2.140}$$

and

$$z^* = \frac{z'}{2\sqrt{\kappa_2 t}} + \frac{(2n+1)h}{2\sqrt{\kappa_2 t}} \qquad I_1(x) = \sqrt{\pi}\ \mathrm{ierfc}\ x. \tag{2.141}$$

In the more complicated case of a Gaussian distribution of intensity in the cross section of the incident laser beam, calculations have been made [2, 3, 102] to different orders of perturbation theory.

2.8. Superficial melting and metal heating beyond the melting point

After reaching the melting temperature on the surface, T_m, and after further dissipation of laser energy into the sample, the melted zone advances into the metal's depth—generally exhibiting thermophysical properties which differ significantly from the thermophysical properties of the solid state metal. One then faces a situation comparable with that discussed in the previous section, i.e. of a metallic sandwich (figure 2.13) which differs from the system metal base plus metal layer by the very fact that in this second case the separation border between the two metallic phases moves in time into the metal.

Assuming that the intensity of the incident laser radiation is constant, $I(t) = I_0$, without taking into consideration, for the time being, the vaporisation and sputtering phenomena, laser heating of the sample up to the melting point T_m and beyond is usually described by the following equation system, with initial and boundary conditions:

$$k_{T_2}\frac{\partial T_2}{\partial z} - k_{T_1}\frac{\partial T_1}{\partial z} = L_m \rho \frac{\mathrm{d}z_m(t)}{\mathrm{d}t} \tag{2.142}$$

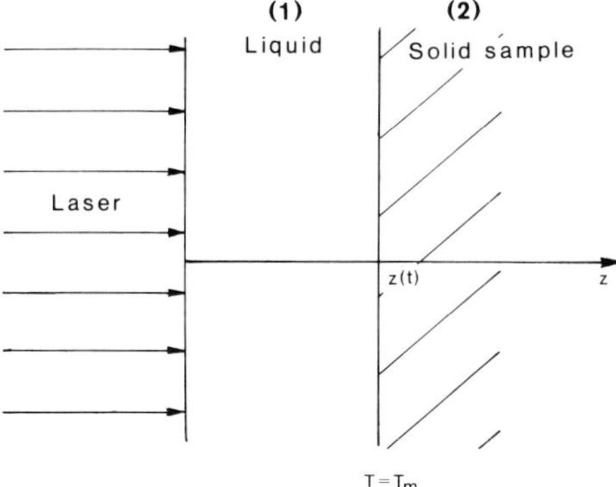

Figure 2.13 Scheme for the laser heating of a solid (metal) in the presence of a liquid phase.

$$-k_{T_1}\frac{\partial T}{\partial z}\bigg|_{z=0} = AI_0 \tag{2.143}$$

$$\frac{\partial^2 T_i}{\partial t^2} = \kappa_i \frac{\partial^2 T_i}{\partial z^2} \qquad i = 1, 2 \tag{2.144}$$

$$T_1 = T_2 = T_m \qquad z = z_m(t) \qquad t > 0 \tag{2.145}$$

$$T_2(z, 0) = T_0 = 0 \tag{2.146}$$

$$T_2(\infty, 0) = T_0 = 0 \tag{2.147}$$

$$z_m(0) = 0. \tag{2.148}$$

Here L_m is the latent heat of melting, indices 1 and 2 specify the liquid layer and the metallic support, respectively, and $z_m(t)$ stands for the solid/molten metal interface.

Details concerning the analytical—or more frequently numerical—solution of the equation system (2.142)–(2.148) are not given here. The problem becomes even more difficult when the temperature dependences of the thermophysical and optical parameters are introduced, and when one attempts to model their variation with temperature in the case of a liquid metal. We will confine ourselves here to some characteristic results and features that are useful in the analysis of melting of metal samples in various situations of practical interest.

Let us start with the determination of the moment t_m at which superficial melting of the metal sample is initiated (taking the beginning of irradiation as the time origin). This can be done by using, as appropriate, one of the many expressions for the temperature field of targets of various shapes, taking in these $T = T_m$ and $z = 0$, that is

$$T(r, 0, t_m) = T_m. \tag{2.149}$$

When the related calculations are difficult, or whenever a rapid evaluation is necessary, one can resort to the simpler and obvious relation

$$t_m \simeq \frac{V}{AE_0}(\rho c T_m + L_m) \tag{2.150}$$

where V is the metal volume where the laser energy is dissipated. For very short laser pulses one can evaluate V from

$$V = S_s \delta \tag{2.151}$$

where $\delta = 1/\alpha$ is the depth of the skin layer, while whenever $l_{th} > \delta$ one can apply the relation (2.43).

In order to evaluate the time t_z necessary for melting a surface layer of the sample of thickness z_m one can use the formula

$$I_0(A^s t_m + A^l t_1) \simeq z_m(\rho c T_m + L_m) \tag{2.152}$$

where t_1 is the time required for melting the whole layer starting from the moment when the melting temperature, T_m, was reached on surface, i.e. we have $t_z = t_m + t_1$; A^s and A^l are the metal absorptivity in solid phase and in liquid phase, respectively.

In practice, it is interesting to determine the thickness of the superficial liquid layer, which can be created and maintained on the surface of metallic targets under the action of power-laser radiation when substance removal in the form of vapours and/or liquid is avoided [103, 104].

The numerical estimations can be performed starting from the formulae in the preceding paragraphs, on the condition that on the sample surface, $z = 0$, the temperature reaches the boiling point, $T = T_v$, and determining the depth, $z = z_m$, at which the temperature is equal to the melting point, $T = T_m$. In the case of a Gaussian space distribution, calculations are made in the centre of the irradiation spot, that is for $r = 0$. Therefore

$$T(0, z_m, t) = T_m \tag{2.153}$$

$$T(0, 0, t) = T_v. \tag{2.154}$$

For large irradiation spots (uniform heating), and assuming a constant absorptivity, one obtains

$$z_m = \frac{1.2\, k_T T_m}{A I_0}\left(\frac{T_v}{T_m} - 1\right). \tag{2.155}$$

For a uniform distribution of energy in a circular spot of radius R_s and long irradiation times ($t \to \infty$), i.e. for three-dimensional heating, we have

$$z_m = \frac{R_s}{2}\left(\frac{T_v}{T_m} - \frac{T_m}{T_v}\right). \tag{2.156}$$

And finally, in order to evaluate the size of the molten area on the irradiation surface, one can start also from the respective expression of the temperature field, and setting the conditions

$$T(r = r_m, 0, t) = T_m \tag{2.157}$$

$$T(0, 0, t) = T_v. \tag{2.158}$$

In other applications instead of $T = T_v$ in equations (2.154) or (2.158) one can use any other temperature $T > T_m$, as appropriate.

Chapter 3 Light-induced Thermoelastic Deformation of Metal Surfaces

Dynamic deformations of metal surfaces due to irradiation-induced reversible thermodeformations and irreversible plastic thermodeformations are discussed in this chapter. It is shown that both can entail strict limitations in the performance of the laser metallic components.

At incident laser radiation intensities insufficient for initiating target vaporisation or melting, one can still notice modifications of the irradiated surface as a result of thermoelastic deformation. We shall examine in some detail these phenomena, which can determine significant limitations in the operation with mirrors—in particular with the metallic mirrors used in conjunction with power-laser systems. The discussion is mainly based upon the results obtained in references [105–108].

The attempt to describe the thermoelastic behaviour of a continuous medium under the action of a pulsed heat source creates a difficult mathematical problem. In order to deal with only simple analytical expressions, having at the same time a general enough character, several assumptions are made. First, we shall consider the sample as semi-infinite and the heat source as a surface source. Secondly, the energy distribution in a cross section of the laser beam will be described with a Gaussian function of the type (2.3), i.e. we shall take $R_s = R_s^e$. Thirdly, in order to have the possibility to perform a separate analysis of the thermal and mechanical sides of the problem—in other words, to be able to ignore the thermal inertia—we shall consider laser pulses having sufficiently long durations

$$\tau_p > R_s/u_0 \tag{3.1}$$

where u_0 is the sound velocity in the solid (in the case of metals $u_0 \sim 10^5\ \mathrm{cm\,s^{-1}}$).

The temperature dependence of optical, thermophysical and mechanical properties of the metal, as well as the convective and radiation energy losses from the target surface are ignored.

To begin with, we shall examine the main phases of the thermoelastic surface deformation process, and then we shall look into the consequences of this phenomenon.

3.1. The main phases of the process

3.1.1. The temperature field

As indicated, the temperature evolution in the irradiation spot is given by equation (2.37) which, taking into account the obvious relation

$$\arctan y = \frac{\pi}{2} - \arctan \frac{1}{y} \tag{3.2}$$

becomes

$$T_G(0,0,t) = \frac{AI_0R_s}{k_T}\left(\frac{\sqrt{\pi}}{2} - \frac{\arctan(R_s^2/4\kappa t)^{1/2}}{\sqrt{\pi}}\right). \tag{3.3}$$

Let us introduce the Fourier number

$$F_0 = \frac{\kappa t}{2R_s^2} \tag{3.4}$$

and the function

$$f_T(F_0) = [\pi - 2\arctan(1/2\sqrt{2F_0})]^{-1} \tag{3.5}$$

With these, equation (3.3) takes a particularly useful form

$$T_G(0,0,t) = \frac{AI_0R_s}{2k_T\sqrt{\pi}}\,\frac{1}{f_T(F_0)}. \tag{3.6}$$

The peculiarity of the temperature function $f_T(F_0)$ (figure 3.1) consists in the fact that, in time (i.e. with increase in F_0) its value quickly decreases and then, with the increase in the Fourier number, it saturates at $f_T(\infty) < 1$. Thus, according to the calculations performed with the aid of equation (3.6) with $F_0 \geqslant 4$, the respective temperature differs from the stationary value by less than 10%.

We also note that, for short irradiation durations, when the temperature rises quickly, the temperature $T_G(0,0,t)$ no longer depends, to any significant degree, on the focal spot size—because the heat losses are insignificant.

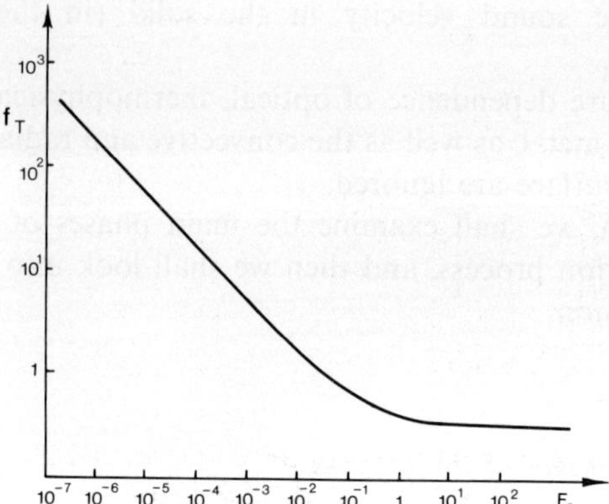

Figure 3.1 The temperature function $f_T(F_0)$.

In fact, for $F_0 \ll 1$ one can consider that a plane source is used for surface heating.

The computed curves $T_G(r, z=0, t)$ of the temperature profiles on an aluminium sample surface at different moments in time, when the sample is submitted to the action of cw CO_2 laser radiation of 1 kW power ($R_s = R_s^e \simeq 7$ mm), are represented in figure 3.2. In addition, figure 3.3 shows the evolution of the stationary temperature field ($t \to \infty$ or for $F_0 \gg 1$), close to the irradiation zone on the surface, $T_G(r, 0, \infty)$ and into the body of the aluminium sample, on its axis, for different R_s.

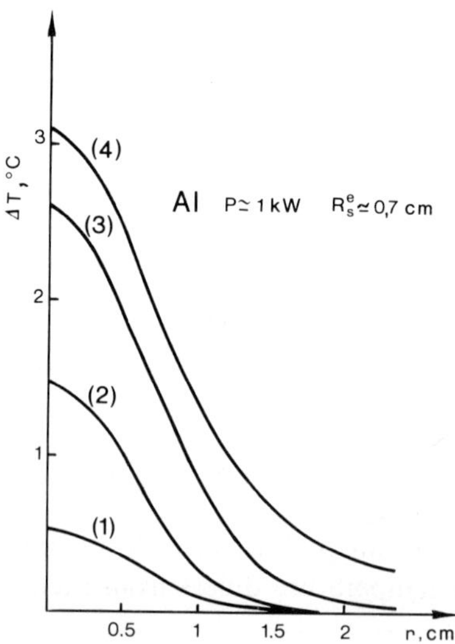

Figure 3.2 Temperature distribution profiles on the surface of a mirror subjected to cw CO_2 laser irradiation during a time $\tau_p = 10^{-2}$ s (1), $\simeq 10^{-1}$ s (2), $\simeq 1$ s (3) and $\simeq 10$ s (4), respectively.

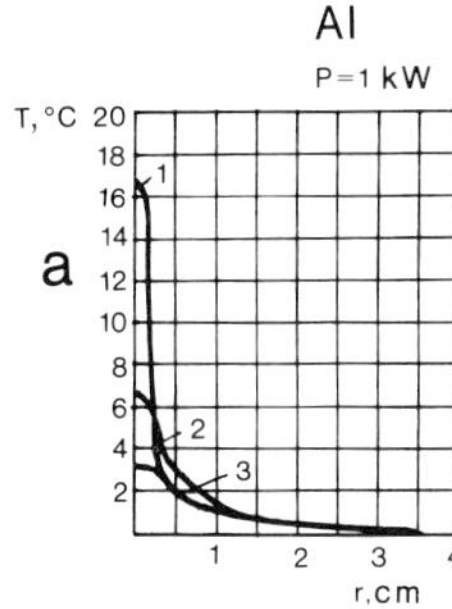

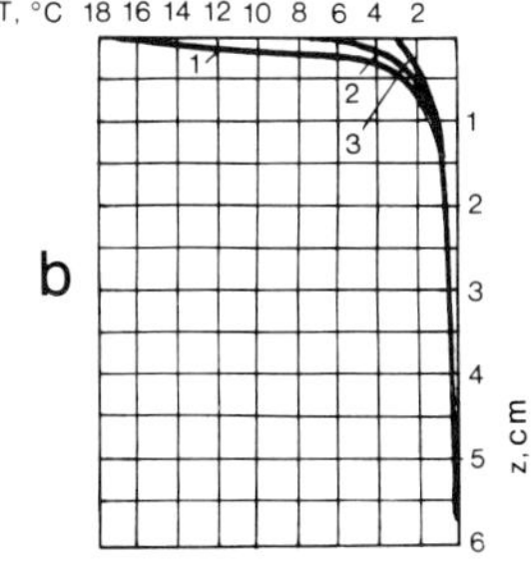

Figure 3.3 The stationary temperature field into an aluminium mirror under power ($P \simeq 1$ kW) CO_2 laser irradiation, at various radii of the focal spot $R_s^e \simeq 1.5$ mm (1), $\simeq 3.5$ mm (2) and $\simeq 7$ mm (3), respectively. (*a*) Surface distribution; (*b*) in-depth distribution.

One notices that the temperature field is localised near the irradiation spot. One must also emphasise the existence of strong temperature gradients, both along the radial direction and into the sample.

3.1.2. *Thermoelastic stresses*

Under the action of fast transitory temperature fields, thermoelastic stresses occur in the metal. In a quasi-stationary approach, the potential $\Delta\varphi$ of the thermoelastic stresses that appear by the action on the surface of a heating source of axial symmetry [109, 110] results from the equation

$$\Delta\varphi \simeq \alpha_T \frac{1+v}{1-v} T \tag{3.7}$$

with the following initial and boundary conditions

$$\varphi(r, z, 0) = 0 \tag{3.8}$$

$$\partial\varphi(r, z, 0)/\partial t = 0. \tag{3.9}$$

Here $T = T_G$ stands for the temperature change, while α_T is the coefficient of linear dilatation and v is Poisson's coefficient, corresponding to the sample material.

To analyse equations (3.7)–(3.9) is no easy task. That is why we shall confine ourselves to a few of the most important conclusions that result from their solution, and that prove useful for the purpose of this work.

First, the solution of equations (3.7)–(3.9) allows for the calculation of some of the stress-tensor components, σ_{ik} (figure 3.4). In the stationary case ($F_0 \gg 1$) and in cylindrical coordinates, the component σ_{rz} along the $0z$ axis is always zero, whereas the most interesting are the radial, σ_{rr}, and azimuthal, $\sigma_{\varphi\varphi}$, components of the tensor which become maximal on the irradiated surface, and they rapidly decrease when progressing into the sample's depth.

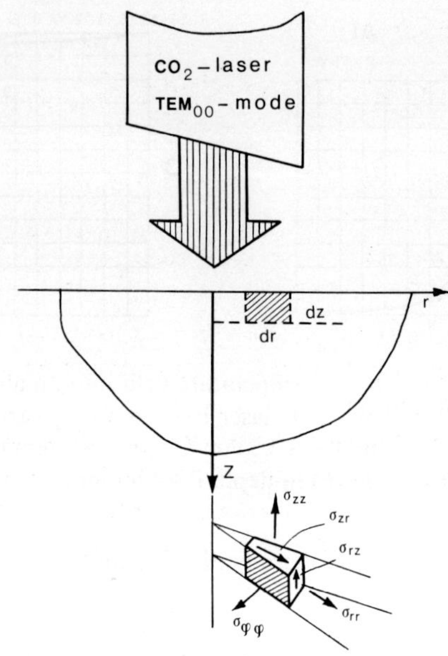

Figure 3.4 Diagram illustrating the thermoelastic behaviour of a continuous medium exposed to a power-laser beam, in the model semi-infinite solid.

Thus

$$\sigma_{rz}(r, 0, \infty) = \sigma_{zz}(r, 0, \infty) = 0 \tag{3.10}$$

$$\sigma_{rr}(r, 0, \infty) = -\frac{\alpha_T A I_0 E_Y R_s^2}{2k_T} \int_0^\infty \exp(-\xi^2 R_s^2/4) \frac{J_1(\xi r)}{\xi r} \, d\xi \tag{3.11}$$

$$\sigma_{\varphi\varphi}(r, 0, \infty) = -\frac{\alpha_T A I_0 E_Y R_s^2}{2k_T} \int_0^\infty \exp(-\xi^2 R_s^2/4)\left(J_0(\xi r) - \frac{J_1(\xi r)}{\xi r} \right) d\xi. \tag{3.12}$$

Here J_0 and J_1 are Bessel functions of the first kind, of zero and first order, and E_Y is Young's modulus.

In the centre of the irradiation spot the only tensor components differing from zero are σ_{rr} and $\sigma_{\varphi\varphi}$, and they have equal values to be expressed by the simple form

$$\sigma_{rr}(0, 0, t) = \sigma_{\varphi\varphi}(0, 0, t) = \frac{\alpha_T E_Y}{2(1-\nu)} T \tag{3.13}$$

which, for $F_0 \ll 1$ becomes

$$\sigma_{rr}(0, 0, t) = \sigma_{\varphi\varphi}(0, 0, t) = \frac{\alpha_T A I_0 E_Y R_s}{2\sqrt{\pi}(1-\nu)k_T} \arctan(4\kappa t/R_s^2)^{1/2} \quad (3.14)$$

while in the stationary case ($t \to \infty$, $F_0 \gg 1$) we obtain

$$\sigma_{rr}(0, 0, \infty) = \sigma_{\varphi\varphi}(0, 0, \infty) = \frac{\sqrt{\pi}}{4} \frac{\alpha_T A I_0 E_Y R_s}{k_T(1-\nu)}. \quad (3.15)$$

The dependence on the coordinate r of the stress tensor components, σ_{rr} and $\sigma_{\varphi\varphi}$ in a stationary case, on the surface of various metal targets, for $AP = 1$ kW and $R_s^e = 1$ cm calculated on the basis of relations (3.11), (3.12) is represented in figure 3.5. We point out the change in sign of the stress component $\sigma_{\varphi\varphi}$ that occurs at a distance r_0 from the spot centre—close to, but a bit larger than the spot of radius R_s. It corresponds to the transition from the material compression in the heating area (negative values of the stress-tensor components) to its expansion outside the spot.

For completeness, figure 3.6 shows, as a typical example, the variation into the sample's depth, of the axial component of the stress tensor, σ_{zz}, in the irradiation spot centre ($r = 0$), for an aluminium sample exposed to cw CO_2 laser irradiation, $P \simeq 1$ kW, of various durations. One can notice that $\sigma_{zz}(0, z, t)$ exhibits a maximum, whose position along the 0z axis, $z = 0$, approaches the surface when the irradiation duration decreases.

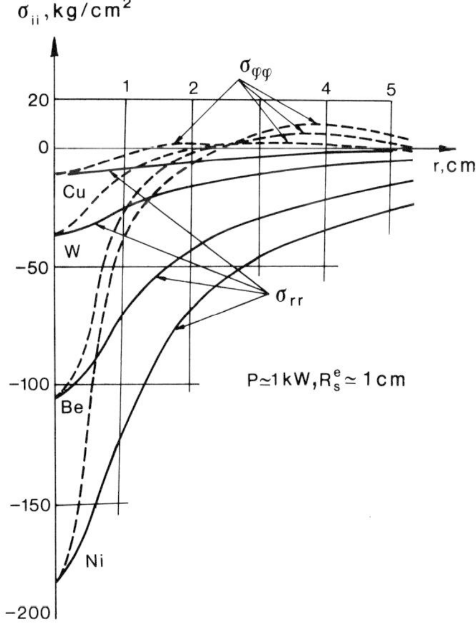

Figure 3.5 The stationary stress field on the irradiated surface of mirrors made of various metals.

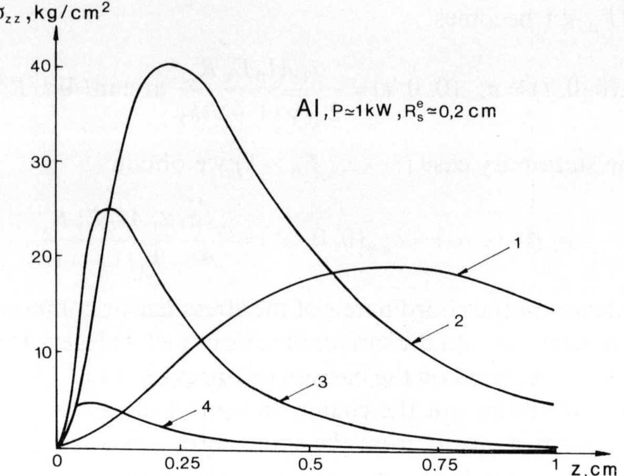

Figure 3.6 Variation of the axial component of the stress tensor, $\sigma_{zz}(0, z, t)$, beneath the centre of the irradiation spot into the depth of an aluminium sample exposed to cw CO_2 laser irradiation, $P = 1$ kW, after a dwell time $\tau_p \simeq 10$ s (1), $\simeq 1$ s (2), $\simeq 10^{-1}$ s (3) and $\simeq 10^{-2}$ s (4).

3.1.3. *Surface deformation*

Thermal stresses cause distortion of the irradiated surface, with a characteristic linear scale $\delta(r, t)$, which increases with sample temperature and σ_{ik} values.

By solving equations (3.6)–(3.9) the following expressions [113] are obtained for the distortions $\delta(r, t)$ in the direction normal to the target surface

$$\delta(r, t) = -(1+\nu)\frac{\alpha_T A I_0 \kappa R_s^2}{k_T}\int_0^\infty \exp(-\xi^2 R_s^2/4) J_0(\xi r)$$

$$\times\left[t + \operatorname{erf}(\xi\sqrt{\kappa t})\left(\frac{1}{2\kappa\xi^2} - t\right) - \frac{\sqrt{t}}{\xi\sqrt{\pi\kappa}}\exp(-\xi^2\kappa t)\right]\xi\, d\xi. \quad (3.16)$$

The thermal deformation profiles of some copper surfaces at different moments in time during laser irradiation are represented in figure 3.7.

One can see that even in the case of relatively low power levels of the radiation absorbed into the metal (e.g. in the case of copper samples with an absorptivity $A = 0.01$ and an absorbed power $AP \simeq 10$ W), the amplitude of the surface deformations (distortion), reaches a few fractions of a micrometre and is a maximum on the radiation axis. It is also emphasised that, once initiated, the process of surface profile distortion can go on indefinitely: indeed, as shown in figure 3.7, there is no saturation in $\delta(r, t)$, even for irradiation durations of $t = 100$ s.

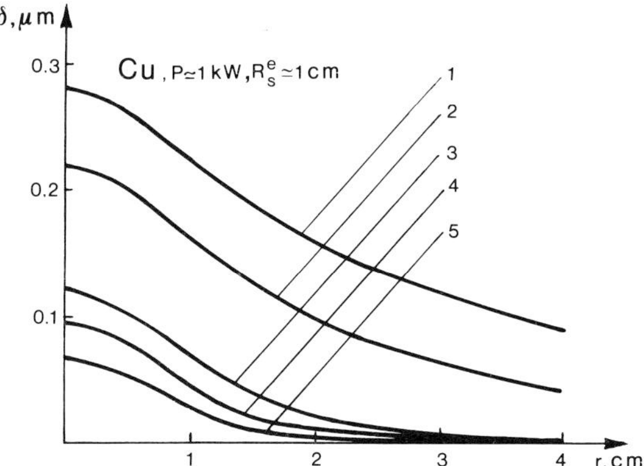

Figure 3.7 Dynamics of the profile modification of an initially plane surface of a copper mirror at different time moments during cw CO_2 laser irradiation $t = 10^2$ s (1), 10 s (2), 1 s (3), 0.5 s (4) and 0.25 s (5).

The time evolution of the amplitude of thermodeformation at the centre of the irradiation spot $\delta(0, t)$ for different materials—metals and others—is represented in figure 3.8. One can see that different materials exhibit a different stability to thermodeformation at different moments in time during irradiation. For example, fused silica with a thin gold deposition on surface (in order to increase reflectivity) presents a lower distortion $\delta(0, t)$ in comparison with other materials for small durations of laser action, $t \leqslant 1$ s. The situation becomes even more complex as τ_p increases. For example, at practically the same AP, the deformation of fused silica by laser irradiation

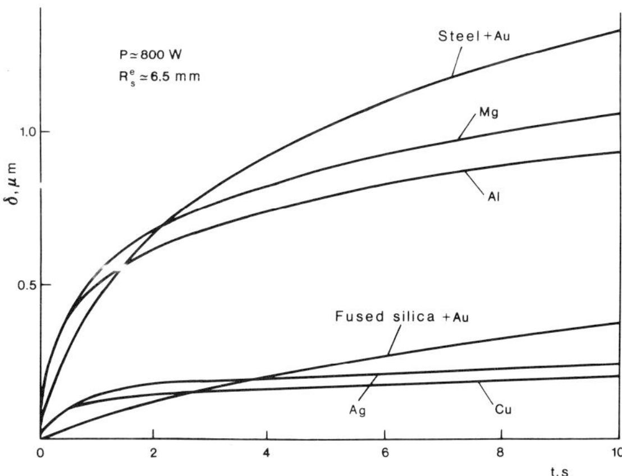

Figure 3.8 The modification in time of the distortion $\delta(0, t)$ in the centre of the irradiation spot, for mirrors manufactured of different materials. The power of the cw CO_2 laser radiation was of $P = 800$ W. $R_s^e = 6.5$ mm.

is larger than in case of irradiation of samples of Ag and Cu, but is less than the thermodeformation of aluminium—a material with very similar thermophysical properties. In order to understand such differences, one has also to take into account in the calculation of $\delta(0, t)$ (besides the thermophysical properties) the elastic properties of the material, in particular the coefficient of linear dilatation. This provision is particularly valid for the following approximations

$$\delta(0, t) = -(1+\nu)\frac{\alpha_T A I_0 R_s^2}{k_T}\frac{1}{f_2(F_0)} \tag{3.17}$$

where

$$f_2(F_0) = \left(4F_0(1-2f_3(F_0)) + \tfrac{1}{4}\ln\frac{1+2f_3(F_0)}{1-2f_3(F_0)} - f_3(F_0)\right)^{-1} \tag{3.18}$$

and

$$f_3(F_0) = \left(\frac{2F_0}{1+8F_0}\right)^{1/2} \tag{3.19}$$

The negative value indicates that distortion appears in the negative sense of the z axis (i.e. towards the laser source). Figure 3.9 shows the computed function $f_2(F_0)$. One can also see that, like function $f_T(F_0)$, the function $f_2(F_0)$ would initially decrease linearly with the increase in F_0, to reach saturation at much higher values of F_0. The rather small decrease in $f_2(F_0)$ for $F_0 \gg 1$ is due to the extension of the heated zone on the sample.

In the limits of small and large irradiation durations, respectively, based upon equations (3.17)–(3.19) one can obtain rather simple expressions

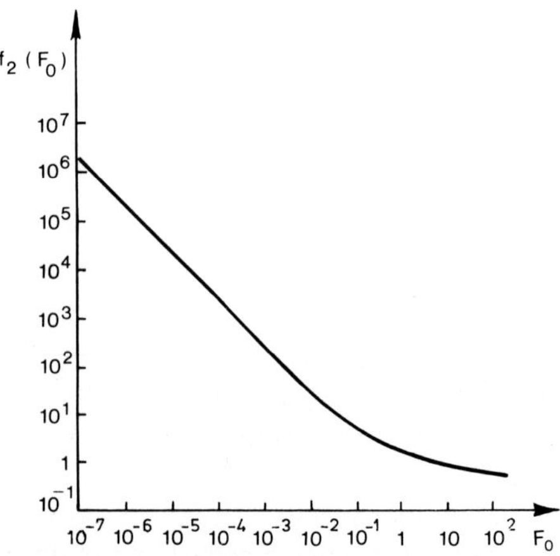

Figure 3.9 The function $f_2(F_0)$.

characterising the thermodeformation dynamics. Thus, one has

$$\delta(0, t) = -2(1+\nu)\frac{\alpha_T A I_0 \kappa t}{k_T} \qquad \text{for } F_0 \ll 1 \tag{3.20}$$

and

$$\delta(0, t) = -(1+\nu)\frac{\alpha_T A P}{4\pi k_T}\ln\left(\frac{16\kappa t}{R_s^2}\right) \qquad \text{for } F_0 \gg 1. \tag{3.21}$$

Several features of metal thermal distortion following from equations (3.16)–(3.21) are worth highlighting, and also provide explanations for the experimental evidence displayed in figure 3.8. First, distortion is in direct proportion to the ratio α_T/k_T. Secondly, thermodeformation increases rapidly at small F_0, whereas at $F_0 \geqslant 1$ the increase is markedly slowed down. For example, in the case of the fused silica samples, the heat diffusivity of which is much lower in comparison with that of metals, the condition $F_0 \geqslant 1$ is reached only at much larger irradiation durations. Thirdly, the value of $\delta(r, t)$ is in direct proportion with the absorbed laser beam intensity AI_0.

3.2. Stability criteria of the surface profile of metallic samples (cw laser irradiation regime)

In many problems of practical interest it is important to know the conditions which allow for keeping the thermodeformation of surfaces below a given level. One such typical problem is to keep the surface of a metallic mirror undistorted over the irradiation process, or, to be more precise, to make sure that the maximum distortion would not exceed a certain given value, δ_{max}. From equation (3.17) one can then obtain the corresponding equation giving the maximum value of the incident intensity to be delivered, $I_{max}^{\delta_{max}}$, for which $\delta(0, t) < \delta_{max}$ [111, 112]

$$I_{max}^{\delta_{max}}(0, t) = \frac{k_T \delta_{max}}{\alpha_T(1+\nu)AR_s^2} f_2(F_0). \tag{3.22}$$

Traditionally, starting from the requirements set for single-mode laser optics, one usually considers as the maximum admissible the distortions in a proportion that does not surpass a value of $\lambda/20$ (λ is the laser radiation wavelength). Accordingly, in equation (3.22) one may take $\delta_{max} = \lambda/20$, and the following expression is established for the corresponding critical intensity of the radiation

$$I_{max}^{\lambda/20}(0, t) = \frac{k_T(\lambda/20)}{\alpha_T(1+\nu)AR_s^2} f_2(F_0). \tag{3.23}$$

We draw attention to the fact that, with high-performance laser systems (when the maximum allowed beam divergence is only $\theta \leqslant 10^{-5}$ rad), the requirements regarding surface irregularities are even more difficult to meet, at a factor $\lambda/100$.

From equations (3.21) and (3.22) it results that the maximum acceptable value of the laser irradiation intensity is higher the greater the thermal conductivity, k_T, and the lower the absorptivity, A, and the linear thermal dilatation coefficient, α_T, of the metal. For the majority of metals, Poisson's coefficient stays within the range $\nu \leqslant 0.5$ and does not influence to any significant extent the quantity $I_{max}^{\lambda/20}$.

With the increase in incident laser radiation intensity, a situation may occur when the induced stresses reach the elasticity threshold σ_m of the target material. In this case, by the time the laser pulse action ceases, irreversible damage of the sample surface is caused. Mention should be made that, for $\sigma_{ik} < \sigma_m$, the surface profile modifications have a reversible character, i.e. when the pulse is over, or the cw laser source is switched off, the surface would cool down, and the mirror surface would not differ from the initial one.

As indicated above, the stresses σ_{rr} and $\sigma_{\varphi\varphi}$ have maximum values on the surface, in the irradiation spot centre. When the value σ_m is reached, the metal flow begins precisely from this zone.

From the stationary solution (2.143) results the following expression for the threshold intensity at which the material flow is triggered:

$$I_{max}^{\sigma_m}(0, \infty) = \frac{4}{\sqrt{\pi}} \frac{k_T \sigma_m}{\alpha_T E_Y A R_s^2}. \tag{3.24}$$

The change in sign of the $\sigma_{\varphi\varphi}$ stress-tensor component appearing at $r_0 = R_s^e$ shows that while the material within the irradiation spot ($r < r_0$) is compressed, the material in the area $r > r_0$ undergoes an expansion. Owing to this behaviour, radial cracks may appear if $\sigma_{\varphi\varphi}(r_0) > \sigma_m$.

Equation (3.24) was obtained in the case of a stationary regime. The analysis is much more complicated in the case of short laser pulses. With the transition to a single- and multiple-pulsed irradiation regime the problem becomes much more complicated and will be examined in detail in Chapter 4, devoted to the resistance of metal mirrors to the effects of pulsed light. Here we only mention that precisely under these regimes the thermoelastic deformation determines in practice the threshold for mirror damage.

Let us examine now the way in which the reversible and irreversible thermodeformation effects determine the limiting light load on metallic mirrors in a cw irradiation regime [105–115].

We first note that thermal deformation of mirrors at $I < I_{max}^{\sigma_m}$ has as an effect the modification during the very process of irradiation of the wavefront of the laser radiation reflected by the mirror surface. This phenomenon has been described in references [30, 38, 80], and it can be analysed through a comparison of the space- and time-shapes of the laser pulse incident on, and respectively reflected by, the sample surface.

On the other hand, for $I_{max}^{\sigma_m} < I_0 < I_m$—where I_m is the intensity threshold which corresponds to the surface melting of the metal—remanent deformations are to be observed, as was experimentally demonstrated by interferometric experiments [105–108].

And finally, metal melting is induced within the irradiation zone for $I_0 > I_m$.

The dependence on the irradiation spot dimension of $I_{max}^{\lambda/20}$ and $I_{max}^{\sigma_m}$ as calculated according to the relations (3.22) and (3.23) for a copper mirror of absorptivity $A = 0.01$, subjected to laser irradiation for a period $\tau_p = 2$ s ($F_0 \gg 1$), is represented in figure 3.10, which also illustrates the evolution of the radiation intensity threshold which corresponds to the onset of the surface melting of the metal, I_m.

From figure 3.10 we note that, in a cw irradiation regime, $I_{max}^{\lambda/20}$ is normally much smaller than $I_{max}^{\sigma_m}$ and I_m. That is why thermodeformation of the surface is one key factor determining mirror quality. A first requirement is thus to choose mirrors of minimal absorptivity. A good material for manufacturing mirrors would also have thermophysical and elastic parameters that would allow for minimum thermodeformations on surfaces. To illustrate, table 3.1 gives the computed values of the surface temperature increase $T_f - T_0$ of mirrors of different materials at the centre of the irradiation spot ($\pi(R_s^e)^2 = 1$ cm^2, $P = 900$ W, $\tau_p = 2$ s, a stationary regime), together with the experimental and calculated values of the distortion δ_{theor} and δ_{exp} [105–108, 111–113]. We note the excellent agreement among data. Similar

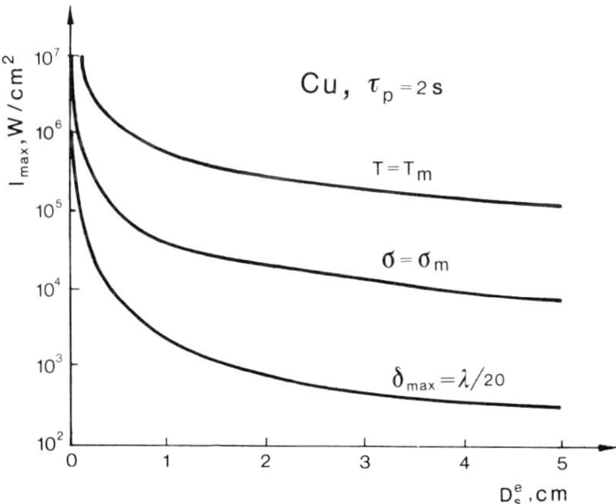

Figure 3.10 Evolution, as a function of the diameter of the incident laser beam, D_s^e ($\lambda = 10.6$ μm), of the threshold values of intensity in the cases of: reversible thermal deformation $\delta_{max} \simeq \lambda/20$; remanent deformation (thermomechanical deformation) $\sigma = \sigma_m$; and surface melting of a copper mirror in a stationary regime.

Table 3.1 Experimental, δ_{exp}, and theoretical, δ_{theor}, values of the thermo-deformation of the surface of metal mirrors at the centre of the irradiation spot ($S_s \simeq 1$ cm^2), at 2 s after the laser ($P = 900$ W) was switched on.

Metal	A	$T - T_0$ (°C)	δ_{exp} (μm)	δ_{theor} (μm)
Ag	0.011	1.2	0.2	0.17
Al	0.018	3.8	0.6	0.63
Be	0.026	8	0.6	0.61
Cu	0.01	1.2	0.5	0.41
Mg	0.015	4.3	0.8	0.7
Mo	0.027	8.4	0.3	0.28
Ni	0.03	14.6	0.8	0.82
Steel + Au	0.013	16	0.7	0.74
W	0.02	4.5	0.1	0.14

data for $I_{max}^{\lambda/20}$, at the same level of P are given in table 3.2. Finally, table 3.3 presents useful constants for the calculation of the thermoelastic surface deformation of a series of materials.

The two afore-mentioned requirements, aimed at minimising the $I_{max}^{\lambda/20}$ parameter of mirrors, may sometimes act in opposite senses. Steel, for example, is known as a material that withstands thermodeformation (particularly when one takes the minimisation of the ratio k_T/α_T as a stability criterion), while exhibiting a rather large absorptivity. Moreover, mirror absorptivity is determined in most cases not only by the intrinsic absorptivity of the metal, A, but mostly by the polishing quality of its surface, which also determines, as a factor of primary importance, the metal's surface hardness. We note, however, that metals as hard as Ni, Be or steel exhibit large values of A. That is why, in practice, one makes recourse to combined mirrors consisting of a metal base of low k_T/α_T ratio and high hardness, coated with a thin layer of metal of low absorptivity A. If one cannot fulfil the requirement for a high hardness of the metal base, then one can use two-layer deposits,

Table 3.2 Threshold values of cw CO_2 laser radiation intensity causing a thermo-deformation of the mirror surface as large as $\lambda/20$.

Metal	R_s (cm)	$T_f - T_0$ (°C)	$I_{max}^{\lambda/20}$ (kW cm^{-2})	
			Exp.	Calc.
Al	0.29	4.5	1.8	2
Be	0.3	13.3	3.7	4
Mg	0.23	4.5	1.9	1.8
Ni	0.14	12.5	4.0	3.9
Steel	0.23	1.0	0.11	0.09

Table 3.3 Thermophysical constants for the calculation of the thermodeformation of metal surfaces.

Metal	α_T (10^{-6} K^{-1})	k_T (cal cm^{-1} °C^{-1})	ρ (g cm^{-3})	c (cal g^{-1} K^{-1})	T_m (°C)	σ_m (10^2 kg cm^{-2})	E_Y (10^5 kg cm^{-2})
Au	14	0.707	19.3	0.031	1063	3.5	8.2
Ag	19	0.974	10.5	0.056	961	3.5	8.1
Al	26	0.504	2.7	0.214	660	5.0	7.2
Mg	26	0.377	1.74	0.250	650	4.0	4.4
Mo	5.4	0.346	10.22	0.066	2610	35.0	33.0
W	4.5	0.480	19.3	0.032	3410	75	41
Be	12	0.350	1.85	0.425	1284	10	25
Cu	17	0.920	8.94	0.092	1083	7	13.2
Steel	14	0.088	7.87	0.107	1500	170	21

Table 3.4 The thresholds of damage by thermoelastic mechanism, in the case of stationary irradiation.

Metal	R_s (cm)	$T_f - T_0$ (°C)	$I_{max}^{\sigma_m}$ (kW cm^{-2})	
			Exp.	Calc.
Al	0.043	53	180	160
Be	0.07	67	61	61
Mg	0.036	71	270	240
Ni	0.1	86	33	30
Steel	0.7	134	0.35	0.13

in which case the thin intermediary layer exhibiting a high hardness is polished until it is coated with the external highly reflecting metal film.

Another approach, already examined in §2.6.3, is related to the reduction of the temperature of the irradiated surface on account of the transition from bulk massive targets to thermally thin metal foils. This allows for heat removal from the irradiation zone not only through the mechanism of thermoconduction into the metal bulk, but also by forced, convective cooling of the opposite surface of the foil. Different methods have been proposed [105, 116, 117] for the manufacture of metal mirrors with forced convective heat evacuation based upon the use of various liquids, including liquid metals as cooling agents. Much emphasis is currently placed on cooling systems that use extended porosity [118]. One mirror assisted by such a cooling system had undergone distortions as small as $\delta \leqslant \lambda/30$ ($\lambda = 10.6\ \mu$m) when exposed to a laser beam ≈ 50 mm in diameter, rated at an average power of 500 W cm^{-2} and a peak power of $\simeq 2$ kW cm^{-2}.

The values $I_{max}^{\sigma_m}$ computed for different metals are given in table 3.4. The comparison of the values $I_{max}^{\lambda/20}$ and $I_{max}^{\sigma_m}$ given in tables 3.2 and 3.4 leads to an important conclusion. Thus, if a certain material shows a large stability of its surface profile it can be rather easily affected by surface thermoelastic damage. And vice versa, as one can easily see, for example, when examining the values of $I_{max}^{\lambda/20}$ and $I_{max}^{\sigma_m}$ for Al and Ni. This is entirely understandable, for the larger the metal hardness the more it is prone to thermoelastic damage.

Finally we stress that, under a stationary regime, the value $I_{max}^{\sigma_m}$ depends rather strongly on R_s ($I_{max}^{\sigma_m} \sim 1/R_s$). As a result, for large irradiation spots, $R_s \geqslant 1$ cm—a case which is particularly interesting when one tries to assess the processing capability of metallic optics—we always have, in a stationary regime, $I_{max}^{\lambda/20} \ll I_{max}^{\sigma_m}$ and even $I_{max}^{\lambda/20} \ll I_m$.

Thus, the maximum permissible light flux values on mirrors used in conjunction with CW power lasers are determined by the reversible modifications of the surface relief, i.e. by the quantity $I_{max}^{\lambda/20}$.

Chapter 4 Light Stability of Metal Mirrors Exposed to Pulsed Irradiation in a Vacuum

This chapter extends the analysis of the optical damage of metals to surface melting, vaporisation and plasma ignition. A comparative discussion is given on the experimental determinations performed by various authors. Stability criteria featuring resistance to laser damage are introduced. Specifics of the damage by multipulse laser irradiation and size effects are discussed, among other types.

The irreversible damage of the surface of metal mirrors leading to the degradation of their optical properties is one essential factor limiting the power output of pulsed laser systems—in contrast to the case of cw lasers, discussed in Chapter 3, where the main limiting factor is the reversible thermoelastic deformation of the mirror surface.

Let us stress the fact that, consistent with the scope of this book, in this chapter only the stability of mirrors to laser light action *in vacuo* is discussed. Indeed, when the interaction evolves in a gaseous environment, a low-threshold optical breakdown plasma may be ignited in front of the irradiated surface, so that damage of the mirror surface will be mainly determined by the action of the breakdown plasma close to the surface rather than directly by the laser beam.

One can identify the following main kinds of surface damage:

(i) thermoelastic modification of the surface relief at $T < T_m$, leading to band slipping, formation of cracks, etc;

(ii) surface melting of local, isolated zones, or even of the whole area of the irradiation spot also accompanied by changes in the surface relief of the mirror;

(iii) metal vaporisation—from distinct local sites as well as from the whole irradiation zone, having as an effect the formation of craters.

The surface of a high-quality mirror has to be, by definition, as free of defects as possible. We shall not pause therefore to discuss the damage thresholds of different types of surface defects and impurities. We shall examine instead several problems of current concern related to the preparation and the cleaning of high-quality laser mirrors.

Speaking of methods of diagnostics and control of the surface damage of laser mirrors, we can list, apart from visual or microscopic inspection of the morphology of the irradiation zone, the following techniques:

(i) inspection of sample's reflectivity, R, either at precisely the same wavelength λ as that of the laser radiation causing the surface damage, or by using a supplementary, probing, low-power laser generating on a wavelength λ_p which can differ from λ:

(ii) recording the electron emission from the irradiation zone, and the determination of the alterations in the extraction work per electron;

(iii) recording the ion emission;

(iv) recording the light sparks during irradiation.

As an illustration, several results are presented in figure 4.1, of measurements of different damage thresholds at pulsed irradiation, E_s^p, for copper samples when the damage is induced by different processes [80]. One can see that, according to the microscopical evidence, only the

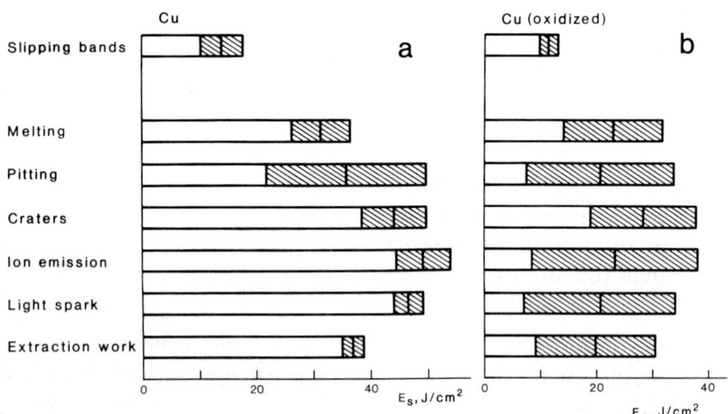

Figure 4.1 The damage thresholds of copper mirrors, with highly processed and carefully cleaned surfaces (a), and after heating in air at 100 °C for a duration of 10 min (b); $\lambda = 10.6$ μm, $\tau_p \simeq 100$ ns.

slip-banding in single-crystal materials is characterised by markedly lower E_s^p values. All the other methods of investigation related to processes determined by the condition of the sample surfaces, give, within the limits of statistical dispersion of data, close values for the damage thresholds.

4.1. Measurement methods

Standardisation of laser damage investigations of optical elements for lasers, including metal mirrors in a vacuum and chemically inert gases, as well as the definition of unitary norms meant to guide the evaluation of the optical stability of optical components such as metal mirrors, can be done only if the accuracy of determination is secured, in each case, for the incident laser energy as well as for its distribution in space and time. Moreover, one has to use radiation with parameters enabling the use, in numerical computations, of the most common and convenient approximations.

For the spatial distribution of the energy, such approximations are the uniform distribution (2.1) and the Gaussian distribution (2.2).

Correspondingly, when a uniform beam intensity, $I(r)=I_0$, $r \leqslant R_s$, is assumed, one can use in threshold determinations either highly multimode laser radiation in combination with appropriate delivery schemes allowing one to select from the beam the zone characterised by the most uniform energy distribution, or one can use a special beam homogeniser, e.g. a fibre-optic waveguide.

Alternatively, one can use monomode laser radiation, having a Gaussian distribution of energy in the focal spot; due to the high stability of the laser parameters, this scheme is mainly used in experiments investigating the laser damage of optical surfaces.

As for the time-shape of the laser pulse, the simplest and most useful approximation assumes a rectangular shape. However, since it is pratically impossible to obtain a perfect rectangular time-shape, one can resort to a bell-shaped pulse, or to a pulse with a leading peak followed by a long-lasting tail, or even to a pulse as close as possible to a rectangular time-shape but presenting less-than-sharp upward and downward slopes.

Let us describe the experimental set-up used in papers [37, 38, 80, 94, 119], that has provided close to ideal parameters for the investigation of the optical stability of optical elements.

The main features of the installation are presented in figure 4.2. Here L indicates the position of the highly reproducible single-mode, single-frequency laser source. The Gaussian energy distribution in a cross section of the beam was tested by automatic sweep of a very fine slit on two perpendicular directions in the focus of the lense F′. Attenuator A was introduced in order to control the laser beam energy; it includes two Brewster

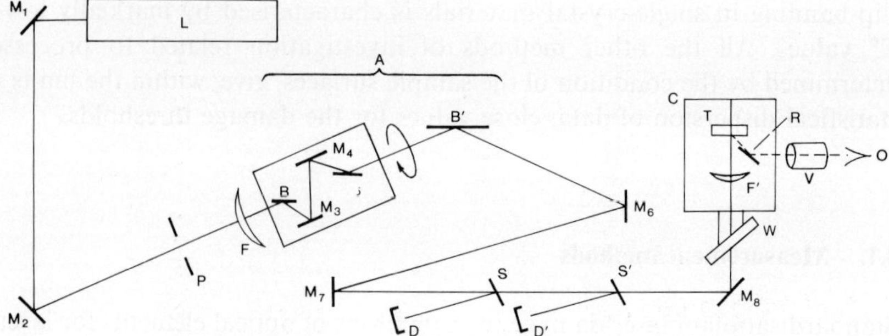

Figure 4.2 Schematic diagram of an experimental installation for measurements of the optical stability of mirrors. The following notations are used: L, laser; M_1–M_8, metal mirrors; B and B′, Brewster plates; P, spatial filter; S and S′, beam-splitters; D and D′, detectors; F and F′, lenses; C, vacuum chamber with W, input window; T, target; V, microscope.

plates, B and B′, in the position of crossed polarisers. Plate B′ is positioned in such a way that it can totally reflect the linearly polarised laser radiation, while plate B can be rotated with a high precision, this being computer assisted. In order not to 'lose' the laser beam as an effect of rotations, and, in general, to ensure reliable operation of the entire installation, the relative position of the copper mirrors M_1–M_8 was computer controlled. Small fractions of the incident beam are deviated by means of the calibrated beam-splitters S and S′, toward a calorimeter, D, and a fast detector, D′, respectively. Their calibration was performed in such a way as to take into account the losses on both the input window W into the irradiation chamber as well as on the focusing lens F′ placed inside the chamber.

In this experiment the laser beam was essentially perpendicular on sample T, placed inside the irradiation chamber. The set-up allowed for the determination of the diffusion reflectivity, R_D, and also—by turning the target—of the specular reflectivity, R_R.

The occurence of a bright spark as well as the modification of the surface profile were observed and recorded *in situ* by means of an optical microscope, V (with a magnifying power $\times$ 20). After irradiation, the exposed area was investigated with a Nomarski microscope.

In the experiments reported in reference [120], the investigation chamber of a scanning electron microscope was used as irradiation chamber (where the laser irradiation of targets was performed (figure 4.3)). This allowed not only a marked improvement in the spatial resolution of surface investigation, but also simultaneous control of the following parameters: the absorptivity of the sample (by a calorimetric method appropriate for thin metal plates); the luminescence of the irradiation zone; and the pulses of electron and ion emission currents.

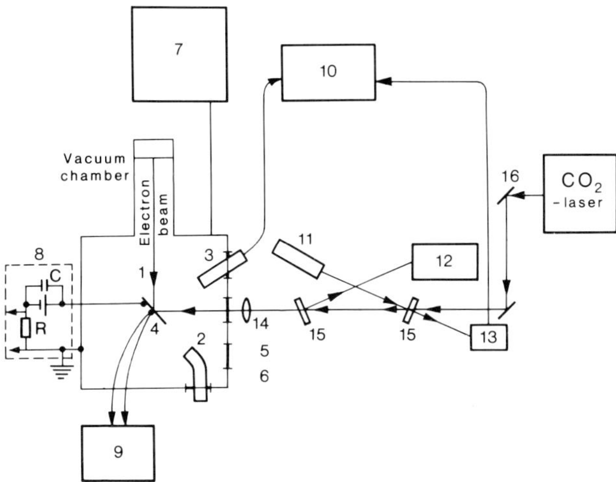

Figure 4.3 Schematic diagram of the diagnostics set-up used for investigating the optical stability of metal optics: 1, target; 2, detector of secondary electrons; 3, fast coaxial detector; 4, thermocouple; 5, Zn–Se window; 6, KCl window; 7, videocontrol device; 8, emission signal recording facility; 9, sampling oscilloscope; 10, oscilloscope; 11, He–Ne laser; 12, calorimeter (energy meter); 13, photodetector; 14, focusing lens; 15, optical delay plates; 16, rotating mirrors.

4.2. Preparation of the mirror surface

We note, first, that the quality of a metal mirror depends on the type of metal, the degree of surface processing, the skills of the manufacturer and the processing facilities available. The last two requirements explain why samples processed in different places, nominally by the same technology, may often show great differences, even when compared with samples prepared by different methods.

In this section we shall review the main technologies in current use for manufacturing laser mirrors. There is, first, the traditional polishing with different abrasive powders. The main disadvantage of this technology derives from the embedding of abrasive particles into the metal base—particularly when soft materials such as Cu and Al, are polished. As an effect, the surfaces of mirrors thus processed present high crystallographic disorder, low thermoconductivity and a large absorptivity. For illustration let us consider the case of a silver mirror whose surface is covered with clusters of particles of a typical diameter $d = 1.5\,\mu\text{m}$, located at distances $l_d \simeq 3$ mm from one another, and which cover about 10% of the sample surface. Assuming an absorptivity $A_d = 2.53 \times 10^{-2}$ for the abrasive particle—a rather small value for these compounds—and taking for silver $A_M = A_D = 3.86 \times 10^{-3}$, we obtain a total absorptivity of the mirror $A = 0.9 \times 3.86 \times 10^{-2} + 0.1 \times 2.53 \times 10^{-2} = 6 \times 10^{-3}$, i.e. $A \simeq 1.5\ A_M$.

Though a certain improvement was made possible by the so-called 'flotation polishing' technique [60], the most significant progress in the manufacture of high-quality mirror surfaces was brought about by a technique consisting of turning the surface with a diamond tool. Particularly good metallic mirrors with surfaces practically free of any defects were obtained by high-speed turning of high-purity metals (for example OFHC—oxygen-free, high-conductivity copper).

We mention that even diamond turning is not entirely free of limitations. These concern the formation, as an effect of the processing, of a relief (spiral), characterised by a period determined by the turning speed. However, in the final analysis the quality of the mirrors obtained by diamond turning is mainly related to the use of a metal base free of any defects.

The homogeneity of a surface layer of metal, of sufficient thickness, prior to diamond turning, can be improved in several ways. One method is surface melting under the action of a scanning electron beam, allowing for a significant reduction in the metal porosity. It is of course much easier to use a thin surface layer than to manufacture the whole mirror from a high-purity base material. That is why several technologies are in use nowadays which are meant to provide metal deposits on metal base plates (by melting, welding, coating, etc), followed by processing by diamond turning.

We also mention that the multilayer structures also open new vistas for abrasive polishing. Thus SiC and Si plates, very pure and practically free of pores and impurities, can be obtained by deposition from a gaseous phase. These exhibit a high thermoconductivity (comparable with or even surpassing the thermoconductivity of copper) and high hardness and can therefore be polished with particular success (without any significant inclusion of abrasive particles into the metal base). After that, a high-purity copper film can be coated onto the polished surface—improving even further the reflectivity of the mirror.

All mirror surfaces, no matter how pure the metal base, and regardless of the sophistication of the manufacturing technology, become dirty in contact with the environment. That is why the surface has to be carefully cleaned before irradiation, e.g. by the following consecutive operations:

(i) degreasing with freons;
(ii) ultrasound washing, in acetone;
(iii) drying in a nitrogen jet;
(iv) repeated washing with acetone just before irradiation and wiping with a fine tissue.

An efficient way of removing adsorbed films or single particles is the direct laser cleaning of the surface. This can be done by scanning the target surface

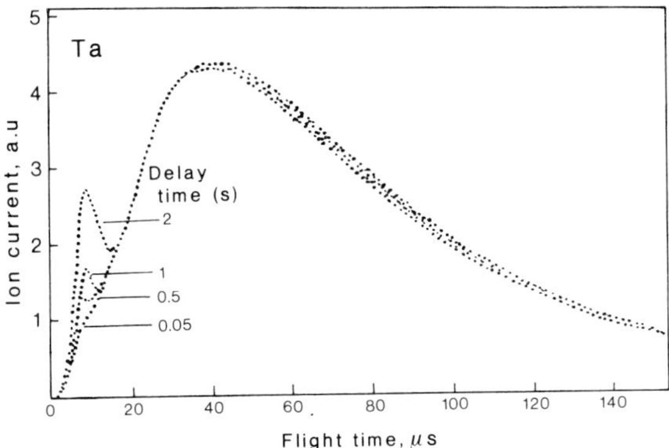

Figure 4.4 The ion current from the surface of a tantalum target for different time delays between two subsequent laser pulses generated by a Nd:YAG laser source. The first pulse with an energy of $E_0 = 17$ mJ and a duration of $\tau_p = 35$ ps performs the surface cleaning while the second ($E_0 = 100$ nJ, $\tau_p = 20$ ns) is used for the vaporisation of and the diagnostic of the material desorbed from the mirror surface. The first peak of the ion current is directly related to the desorbed impurity substance.

with the beam of an additional, CW power-laser source. Another possibility consists in submitting the mirror—just before the investigations meant to test optical stability—to the action of a pulse series, at an incident radiation intensity a little lower than the damage threshold. Unfortunately, any cleaning of the metal surface, including laser cleaning, resists pollution for only a limited time. This is evident from figure 4.4 [64], reporting on targets submitted to the action of two consecutive laser pulses following each other with a controlled delay τ_D. The amplitude of the signal originating in the substance once adsorbed, and removed from the surface, was recorded. One can see that the peak of the impurity signal is absent only for small values of $\tau_D \leqslant 0.05$ s. In other words, in the conditions adopted for the experiments reported in reference [64], the pollution of the surface reappears rather rapidly after laser cleaning. Of course, the later the adsorbed impurities return to the surface, the longer the conditions last which favour the investigation of stability of very pure optical surfaces to the action of light. It would obviously mean higher values for τ_D, and a higher vacuum in the irradiation chamber.

4.3. Thermoelastic damage to the surface

In Chapter 3 we examined the thermoelastic effects in a stationary irradiation regime, when the Fourier number is much larger than unity, $F_0 \gg 1$. In this section, based upon results reported in references [17, 122], we approach the question of thermoelastic deformation and surface damage of metallic mirrors as an effect of pulsed irradiation.

From equations (3.15) and (3.20), and for $F_0 \ll 1$, with a shortening of the duration τ_p of the laser pulse the values $\sigma_{rr}(0, 0, t) = \sigma_{\varphi\varphi}(0, 0, t)$, and by the end of the laser pulse, the distortions $\delta(0, t)$ are reduced. Moreover, at short irradiation durations a non-zero component σ_{rz} of the stress tensor is at work on the surface.

The shift from reversible changes in the surface profile to a process of stress enhancement into the metal—stresses that, beyond a certain threshold σ_m, may cause irreversible damage—is one main feature in characterising the effects on metallic mirrors of pulsed laser irradiation. In addition, pulsed irradiation may cause situations when $I_{max}^{\sigma_m} < I_{max}^{\lambda/20}$, since σ_m can get lower due to cumulative effects under repeated irradiation.

In the following we examine in some detail the basic processes governing the damage mechanisms in metals, and the σ_m values for two different irradiation regimes.

4.3.1. Short monopulse laser action

It is well known that, according to the kind of material, metal damage can occur either by plastic deformation (Al, Cu, etc), or by fragmentation (SiC, Mo, etc). More often than not, damage of metal optics is by plastic flow—evolving as an effect of large plastic deformations. Consequently, one has $\sigma_m = \sigma_T$, where σ_T is the flow limit of the metal. For fragmentation damage one has $\sigma_m = \sigma_B$, where σ_B is the strength modulus.

Figure 4.5 presents the computed curves for plastic damage of aluminium samples (the threshold intensity values $I_p^{\sigma_T}$ were obtained with equation (3.15)

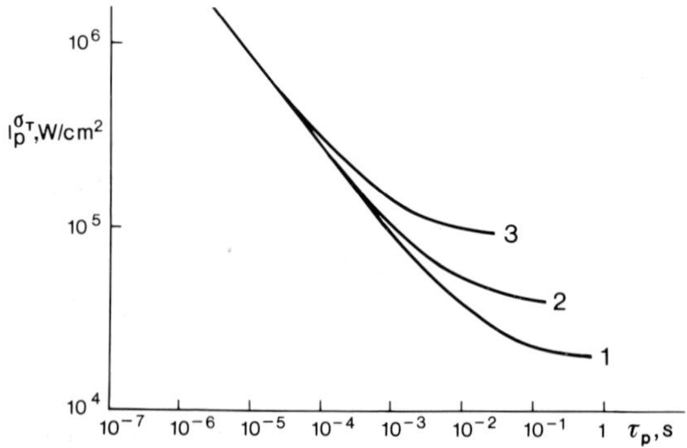

Figure 4.5 The dependence of the stability of the surface to plastic deformation for an aluminium mirror, over the duration of the pulsed action of CO_2 laser radiation, for different irradiation spot sizes, $R_s^e = 1.5$ mm (1), 3.5 mm (2) and 7 mm (3), respectively.

under the assumption $\sigma_T = \sigma_{rr}(0, 0, t) = \sigma_{\varphi\varphi}(0, 0, t)$). One can see that, for $F_0 \ll 1$ (in the case in question for $\tau_p \ll 10^{-2}$–10^{-1} s), the threshold intensity $I_p^{\sigma_T}$ increases linearly with shortening of the laser pulse duration. We note, for example, that for a duration $\tau_p \leqslant 100$ ns, the threshold value is $I_p^{\sigma_T} \geqslant 10^8$ W cm^{-2}. For $F_0 \ll 1$, curves $I_p^{\sigma_T}(\tau_p)$ no longer depend on R_s (this also follows from equation (3.15)).

As already mentioned, one feature of the pulsed damage regime is that the component σ_{rz} of the stress tensor differs from zero. Correspondingly, whenever a sample deformation σ_{zz} is produced under laser irradiation, an expansion appears if $\sigma_{zz} > 0$, and a compression is observed if $\sigma_{rz} < 0$. σ_{zz} reaches its maximum along the 0z axis in a point of coordinates $(0, 0.66R_s^e)$ independent of τ_p, while the site where the quantity σ_{rz} reaches its maximum has the coordinates $(R_s^e/2, \sqrt{2\kappa\tau_p})$.

One can neglect the quantity σ_{zz} on the sample surface. Then the threshold of pulsed surface damage is determined, if irradiation is performed with a Gaussian beam, by two characteristic intensities, derived from the conditions $\sigma_{rr} = \sigma_m$ and $\sigma_{rz} = \sigma_m$ [108, 115]

$$I_p^{\sigma_m}(\sigma_{rr}) = \frac{\sqrt{\pi}}{2\sqrt{3}} \frac{(1-\nu)k_T\sigma_m}{AG\,\alpha_T(1+\nu)} (\kappa\tau_p)^{-1/2} \tag{4.1}$$

$$I_p^{\sigma_m}(\sigma_{rz}) = \frac{\sqrt{2e}}{2} \frac{(1-\nu)k_T\sigma_m}{AG\,\alpha_T(1+\nu)} R_s(\kappa\tau_p)^{-1} \tag{4.2}$$

where, in the case of a monopulse, $\sigma_m = \sigma_T$ or $\sigma_m = \sigma_B$, and $G = E_Y/2(1-\nu)$ is the displacement modulus.

From expressions (4.1) and (4.2) one can infer two conclusions of practical consequence: first, that threshold intensities increase with the shortening of the laser pulse duration τ_p, even though the dependence of the $I_p^{\sigma_m}$ threshold on τ_p differs for the two components of the stress tensor; and second, that the quantity $I_p^{\sigma_m}(\sigma_{rr})$ does not depend on the radius of the irradiation spot, whereas, in contrast, $I_p^{\sigma_m}(\sigma_{rz}) \sim R_s$. This effect is a consequence of the fact that the stress-tensor components σ_{rr} and $\sigma_{\varphi\varphi}$ (as well as, generally, the σ_{zz} component causing damage into the metal's bulk) depend on the distortion gradient along the 0z axis, while for the quantity σ_{rz} we have

$$\sigma_{rz} = G\frac{\mathrm{d}\delta}{\mathrm{d}r}. \tag{4.3}$$

On the other hand, in the one-dimensional approximation ($F_0 \ll 1$), in case of a uniform distribution of energy within the irradiation spot, as described by equation (2.1), $\mathrm{d}\delta/\mathrm{d}r$ and, correspondingly, σ_{rz} are equal to zero. As a result, the mirror damage can be related in this case to the σ_{rr} component of the stress tensor.

For comparison, another expression was introduced for the maximum allowed radiation intensity $I_{\max}^{\lambda/20}$ under the action of which the laser-induced thermal distortion in the centre of the irradiation spot reaches, at the end of the laser pulse, a value $\delta_{\max} = \lambda/20$ [108]

$$I_{\max}^{\lambda/20}(\tau_p) = \frac{\lambda k_T}{40(1+\nu)A\alpha_T}(\kappa\tau_p)^{-1} \quad \text{for } F_0 \ll 1. \tag{4.4}$$

The flow under stress in metals is related to the liberation and motion of dislocations. The triggering of plastic deformation is determined by various parameters (such as the temperature, the growth rate of σ_{ik}, etc). The dependence of σ_T on the dimension of the polycrystalline grains exerts the most significant influence on the quality of the metal optics concerning the stability to thermoelastic damage. Experimental investigations have shown that

$$\sigma_T = \sigma_f + \frac{\text{const}}{(d_g)^{1/2}} \tag{4.5}$$

where σ_f is the friction stress and d_g is the grain dimension. The use of microdispersed polycrystalline materials therefore allows for significantly raising σ_T and consequently for improving the optical stability of mirrors. This is one of the reasons why the intermetallic compounds [117] nowadays find even larger applications in power metal optics—particularly as coatings—either reflective, or inserted between the reflective layer and the metal base.

From the mechanics of damage one knows that for stresses preceding the macroflowing $\sigma_{ik} > 0.96\sigma_T$, microflowing appears, causing a deformation whose relative value can reach $\delta/R_s \leqslant 10^{-5}$.

Parameters σ_T and σ_B diminish during metal heating, tending to zero near melting. Experimental investigations (see, for example, [37, 38, 80, 94]) indicate that the initiation of slip-banding is directly related to the occurrence of the thermoelastic damage, this also being the damage process which features the lowest threshold for metallic surfaces. Figure 4.6 shows a typical image of slip-banding around the melted metal within the irradiation spot. This is also evident from the data for $E_s^p = I_p\tau_p$ represented in figure 4.1 for copper mirrors as well as in table 4.1, where the threshold radiation fluences are given for two laser pulses differing in duration, τ_p, though focused in spots of practically identical radius, R_s, as corresponding to slip-banding, $E_s^p(\sigma_T)$, melting, $E_s^p(T_m)$, vaporisation $E_s^p(T_v)$, and plasma formation, E_s^{pl} for three metals, in vacuum.

Table 4.1 shows that $E_s^p(\sigma_T)$ is two to three times smaller than all the other damage thresholds, and also that all damage thresholds increase with increasing laser pulse duration. In references [94, 119] it was pointed out that, considering the low-threshold thermoelastic damage of surfaces, the

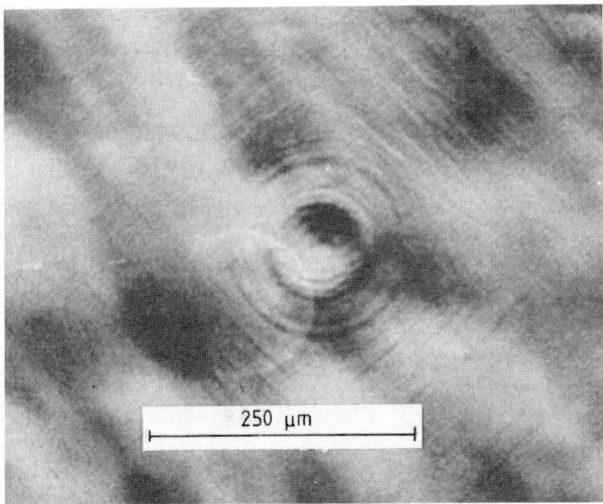

Figure 4.6 Micrograph of slip-banding on the surface of a metallic mirror exposed to pulsed laser irradiation.

choice of the most appropriate material for manufacturing metallic mirrors should start not from the material's melting temperature, but rather from its thermoelastic properties—particularly in the multipulse irradiation regime.

4.3.2. Multipulse irradiation

Specific to multipulse irradiation is the fact that macroscopical damage of the surface is always preceded by microdamage—characterised by the formation and development of microcracks. We note that microcracking results in a worsening of mirror performance, even in the case of monopulse irradiation. This effect is even more important with multipulse irradiation. Indeed, microcracks affect the amplitude and phasefront of the reflected laser

Table 4.1 Experimental damage thresholds of mirrors [94] under the action of CO_2 laser pulses.

Laser pulse	$2R_s$ (μm)	Damage type	E_s^P (J cm^{-2})		
			Cu	Ag	Au
Long	238	Slip-banding	176 ± 4	99 ± 4	152 ± 12
		Melting	476 ± 4	373 ± 7	276 ± 9
		Crater formation	476 ± 4	392 ± 5	280 ± 36
		Plasma ignition	486 ± 6	392 ± 5	276 ± 9
Short	242	Slip-banding	26.4 ± 3.5	19.0 ± 1.9	20.4 ± 3.5
		Melting	69.8 ± 2.1	58.8 ± 1.2	43.4 ± 0.8
		Crater formation	79.0 ± 3.9	42.0 ± 0.9	52.0 ± 0.9
		Plasma ignition	68.8 ± 6.1	72.2 ± 3.7	52.0 ± 0.4

pulse; moreover, their occurrence diminishes plasma ignition thresholds in gases, near mirror surfaces, as a result of the enhanced adsorption of impurities into the microcracks.

From a microscopical point of view, and regardless of the type of damage (be it plastic or fragmentation) the microcracking is initiated beyond a certain stress value $\sigma_l < \sigma_T$, for a uniaxial stress. According to the modified Griffiths theory, the σ_l value is to be obtained for both slipping and shearing from the equation

$$\sigma_l = (\zeta_p E_Y / l_c)^{1/2} \tag{4.6}$$

where ζ_p is the energy of plastic deformation of the surface and l_c the length of the crack.

The microcracking process proceeds at much lower levels of radiation intensity compared with the values initiating macrocracks and damage. The explanation for this is based upon the fact that microcracking is caused by the high concentration of extensile stresses acting upon clusters of dislocations in front of any discontinuity, as for example the mirror surface. In order to avoid the formation of microcracks one has to eliminate, during the mirror manufacturing process, the technological operations that may lead to the generation of dislocations.

Besides, dislocations appear and multiply precisely during the process of cyclic heat loading. This phenomenon leads to so-called fatigue damage, which would manifest itself following a series of many consecutive laser pulses even at levels $\sigma_y \ll \sigma_T$.

Fatigue is a result of the displacement and interaction of the defects in the metal. The damage appears through cumulation of inelastic distortions of the crystalline lattice beyond a critical value, and the destruction of the interatom connections over excessively strained volumes.

The number of cycles after which the amplitude of the fatigue stress tends to its limiting value σ_y^{min} is usually rather high: $N \sim 10^5$–10^7 pulses if the metal's temperature remains lower than a critical value $T_{cr} \sim 500$–$700\,°C$. We have in this case $\sigma_y^{min} \simeq a\sigma_T$, with $a \sim 0.1$–0.3. Alternatively, for temperatures $T > T_{cr}$ and large N, $\sigma_y^{min} \to 0$.

In the case of periodically pulsed irradiation and $\sigma_{ik} > \sigma_y(N)$ the formation of resistant, stable, band slips starts from the beginning (for $N = 1$, the condition $\sigma_{ik} \geqslant \sigma_T$ has to be fulfilled to the same effect), and submicrometre cracks form. Deep bands which widen by further loading would form microcracks whenever band slipping occurs outside the limits of a grain. What results is a catastrophic damage (from the operational point of view) of the mirror surface.

According to reference [112] the maximum allowed intensity of the periodically pulsed radiation $I_p^{pp}(\sigma_{ik})$ can be determined from the equation

$$(I_p^{pp}(\sigma_{ik}))^{-1} = (\tau_p f I_{max}^{\sigma_m})^{-1} + (I_p^{\sigma_m}(\sigma_{ik}))^{-1}. \tag{4.7}$$

Here f is the repetition rate of the consecutive laser pulses. The damage thresholds of the mirror when submitted to a cw irradiation regime, $I^{\sigma_m}_{max}$, and to a monopulse irradiation regime, $I^{\sigma_m}_{p}$, respectively, for $\sigma_m = \sigma_y$, are to be obtained from equations (3.24), (4.1) and (4.2).

One can see from equation (4.7) that the thermoelastic damage of a sample under the action of periodically pulsed radiation is the cumulative result of the action of a continuous component of intensity $I_0 \tau f$—where I_0 is the maximum intensity in a separate pulse—on the background of which temperature peaks and thermoelastic stresses are superimposed, of the same shape as in the case of monopulse action.

4.4. Surface melting of metals

When a metal is irradiated at radiation intensities $I_0 \geqslant I_m$, melting is induced within the irradiation zone. Depending on material purity and the quality of surface processing, two types of surface damage can be identified by $I_0 \geqslant I_m$. For a mirror surface not free from defects, the local pulsed melting of the surface takes place at lower levels of intensity, precisely in the zone where defects are scattered around. This behaviour is clearly indicated in figure 4.7. However, when mirrors of higher quality are used, the melting starts simultaneously everywhere inside the irradiation spot, when the corresponding intensity threshold is reached, $I_0 \simeq I_m$. Such a uniform melting over the whole irradiation spot was reported in reference [94]; it was obtained by laser irradiation of mirrors manufactured from single-crystal copper. The samples were first oriented along one of the crystallographic axes, then submitted to electron polishing and to cleaning in vacuum by sputtering with

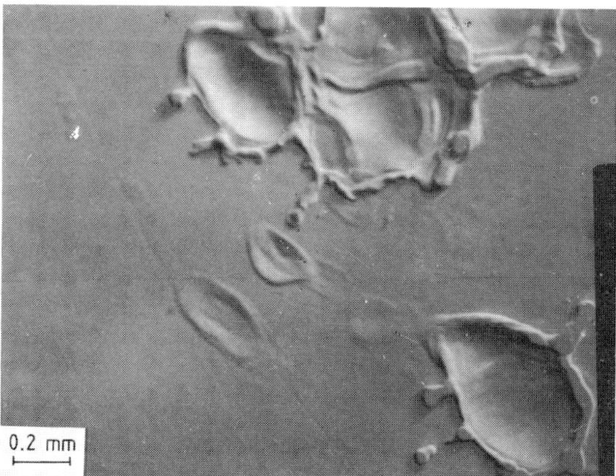

Figure 4.7 Laser-induced local melting of the surface of an aluminium mirror.

an argon ion beam. The remaining argon was then removed by annealing for one hour in vacuum at $T = 500\,°C$. The high degree of surface cleaning, enabling investigations on the sample's resistance to the action of laser light, in particular the determination of the melting threshold, was confirmed by Auger spectroscopy and surface analysis by diffraction with low-energy electrons. All these precautions allowed for finally eliminating any local superheating of the irradiated surface.

To calculate I_m or the corresponding radiation fluence, E_s^m, for metal melting in a surface layer, one usually applies one of the following methods. For the sake of generality, we shall take in all cases the sample as semi-infinite, and will assume that there are large irradiation spots (i.c. $R_s^2 \gg \kappa\tau_p$).

According to the first method—perhaps the most difficult to handle but very accurate indeed—the threshold fluence E_s^m is determined by taking into account the temperature dependence of the absorptivity, $A(T) = A_0 + A_1 T$. By introducing in equation (2.83), $T(0, 0, \tau_p) = T_m$, one obtains the following equation

$$1 + \frac{T_m - T_0}{1 + A_0/A_1} = \exp u_m^2 (1 + \operatorname{erf} u_m). \tag{4.8}$$

By solving equation (4.8) one finds the threshold value $E_s^m = I_m \tau_p$ as

$$E_s^m = \frac{u_m (\tau_p c\rho k_T)^{1/2}}{A_1}. \tag{4.9}$$

For sufficiently small irradiation spots and long irradiation durations, the one-dimensional heat conduction equation that has been used in inferring the solution (4.9) is to be replaced by a three-dimensional equation. A corrected result can thus be obtained [94, 121], for a Gaussian energy distribution in the focal spot, by multiplying the previously obtained value E_s^m by the quantity

$$f_B \simeq \frac{B}{\arctan B} \tag{4.10}$$

where, for $R_s = R_s^e$, we have

$$B = \sqrt{8} (\kappa\tau_p)^{1/2} R_s^{-1}. \tag{4.11}$$

A second, more trivial, approach consists in introducing $T(0, 0, \tau_p) = T_m$ in equation (2.8) and assuming a constant absorptivity, $A(T) = A(T_0)$

$$E_s^{m'} = (\pi c\rho k_T \tau_p)^{1/2} (T_m - T_0). \tag{4.12}$$

And finally, another threshold value of the incident laser fluence required to melt the surface is obtained from

$$E_s^{m''} = \frac{(\pi c\rho k_T \tau_p)^{1/2} (T_m - T_0)}{2A_{md}} \tag{4.13}$$

with

$$A_{\mathrm{md}} = \frac{A(T_0) + A(T_{\mathrm{m}})}{2}. \tag{4.14}$$

The computed values $E_{\mathrm{s}}^{\mathrm{m}}$, $E_{\mathrm{s}}^{\mathrm{m}'}$ and $E_{\mathrm{s}}^{\mathrm{m}''}$ are given in table 4.2 for four metals and different durations of incident laser pulse. The computations were performed based upon cold absorptivity values (i.e. initial absorptivity values) resulting from Drude's theory, i.e. we have assumed $A(T_0) = A_{\mathrm{D}}(T_0)$; some other, typical, experimental values for $A(T_0)$ were also considered. In all cases, A_1 was considered to be equal to the temperature coefficient from the Drude model.

From table 4.2 we notice that the computed thresholds $E_{\mathrm{s}}^{\mathrm{m}}$ and $E_{\mathrm{s}}^{\mathrm{m}''}$ are close to each other for all values of τ_{p} and $A(T_0)$, while the threshold $E_{\mathrm{s}}^{\mathrm{m}'}$—obtained without taking into account the $A(T)$ dependence—clearly exceeds the other melting thresholds for metal mirrors.

The values of melting thresholds are only in good agreement with the experimental findings on the melting thresholds of metal mirror surfaces for samples with very clean surfaces, when one can use the tabulated values of the optical characteristics of metals. For example [80, 94], when samples of single-crystal copper were used in investigations, surface melting of the metal was uniformly induced over the whole irradiation spot at an incident laser fluence of $76.6 \pm 8.6\ \mathrm{J\,cm^{-2}}$, while the computations give the melting threshold as $76.1\ \mathrm{J\,cm^{-2}}$.

In other cases one notices a good qualitative agreement between the experimental and the computed data. On this line, a square-root dependence on duration $E_{\mathrm{s}}^{\mathrm{m}} \sim \sqrt{\tau_{\mathrm{p}}}$ has been confirmed experimentally [37, 38, 80, 94, 119].

Several authors [122–124] have investigated the angular dependence of the surface melting thresholds under the action of linearly polarised laser radiation. When assuming $A(T) = A(T_0) = \mathrm{const}$, one can obtain from equation (4.12) the expressions wherefrom the threshold values $E_{\mathrm{s}}^{\mathrm{m}'}$ can be computed in the case of incident laser radiation linearly polarised parallel ($\|$) and perpendicular ($\perp$), to the incidence plane, respectively

$$(E_{\mathrm{s}}^{\mathrm{m}'}(\theta))_{\|} = \frac{T_{\mathrm{m}} - T_0}{2A(T_0)} (\pi c \rho k_{\mathrm{T}} \tau_{\mathrm{p}})^{1/2} \tag{4.15}$$

and

$$(E_{\mathrm{s}}^{\mathrm{m}'}(\theta))_{\perp} = \frac{T_{\mathrm{m}} - T_0}{2A(T_0)\cos^2\theta} (\pi c \rho k_{\mathrm{T}} \tau_{\mathrm{p}})^{1/2}. \tag{4.16}$$

From equations (4.15) and (4.16) one notices that, while an oblique incidence has no effect in the case of the radiation polarised parallel to the incidence

Table 4.2 Computed values of incident laser fluences for melting the surface of metal mirrors.

$A(T_0) \times 10^{-2}$	E_s^m (J cm^{-2})					$E_s^{m'}$ (J cm^{-2})					$E_s^{m''}$ (J cm^{-2})				
	$\tau_p = 1$ ns	100 ns	1 μs	10 μs	100 μs	$\tau_p = 1$ ns	100 ns	1 μs	10 μs	100 μs	$\tau_p = 1$ ns	100 ns	1 μs	10 μs	100 μs
Ag 3.86 (A_D)	6.08	60.8	192	608	1920	21.9	219	692	2190	6920	7.19	71.9	227	719	2270
4.7	5.22	52.2	165	522	1650	18.1	181	574	1810	5740	6.46	64.6	204	646	2040
5.0	5.09	50.9	161	509	1610	16.9	169	535	1690	5350	6.29	52.9	199	629	1990
7.92	4.25	42.5	134	425	1340	10.8	107	338	1070	3380	5.17	51.7	163	517	1630
Al 10.6 (A_D)	1.79	17.9	56.8	179	567	4.07	40.7	129	407	1290	2.08	20.8	67.0	208	670
12.4	1.78	17.8	56.4	178	564	3.48	34.8	110	348	1100	2.02	20.2	64.0	202	645
13.0	1.73	17.3	54.9	173	549	3.32	33.2	105	332	1050	1.97	19.7	62.2	197	622
18.2	1.42	14.2	44.8	142	448	2.37	23.7	75	237	750	1.59	15.9	50.3	159	503
Au 6.48 (A_D)	4.04	40.42	128	404	1280	13.9	139.5	441	1395	4410	4.79	47.9	151.5	479	1515
7.80	3.46	34.6	109	346	1090	10.55	105.5	333.5	1055	3335	4.11	41.1	130	411	1313
7.87	3.43	34.3	108	343	1085	10.46	104.5	331	1045	3310	4.075	40.75	129	407.5	1290
7.15	3.67	36.7	116	367	1160	11.5	115	364	1150	3640	4.41	44.1	140	441	1400
Cu 4.94 (A_D)	5.74	57.4	182.5	574	1825	22.4	224	707	2240	7070	6.76	67.6	214	676	2140
7.5	4.59	45.9	145	459	1450	14.8	148	467	1475	4670	5.65	56.5	179	565	1790
7.7	4.55	45.5	144	455	1440	14.4	144	454	1440	4540	5.59	55.9	177	559	1770
11.0	3.94	39.4	125	394	1250	10.1	101	318	1010	3180	4.79	47.9	151	479	1510
11.4	3.88	38.8	123	388	1230	9.71	97.1	307	971	3070	4.71	47.1	149	471	1490
13.3	2.75	27.5	87	275	870	8.33	83.3	263	833	2630	3.26	32.6	103	326	1030

plane, it results in a significant increase in the maximum allowed (threshold) optical load onto the sample surface in case of a radiation polarised perpendicular to the incidence plane. The experiments reported in references [123, 124] and performed with linearly polarised CO_2 laser radiation pulsed in portions of $\tau_p \simeq 1.7$ ns have confirmed the dependence $E_s^p(\theta) \sim 1/\cos^2\theta$ as following from equation (4.10), in the case where the radiation was polarised perpendicular to the incidence plane. Samples prepared from different metals were used in these experiments. The largest amplification in the $E_s^p(\theta)$ value was observed for steel mirrors: from 0.6 J cm^{-2} at perpendicular incidence ($\theta = 0°$) to 24 J cm^{-2} at $\theta = 80°$.

Let us finally introduce a quality parameter, as proposed in reference [125] for characterising mirror resistance to surface melting:

$$F_G = \frac{(T_m - T_0)(\rho c k_T)^{1/2}}{A(T_0)}. \tag{4.17}$$

The larger F_G, the higher is the intensity of the fluence of the incident radiation which the metallic surface can withstand without melting.

The F_G values for several metals are given in table 4.3, along with the computed threshold fluences for melting, $E_s^{m'}$, calculated with the same values of $A(T_0)$— denoting the cold (room temperature) absorptivity of these metals.

One can notice that Cu and Ag are the metals most resistant to melting when exposed to CO_2 laser radiation. The situation would radically change when passing to radiation with wavelength $\lambda = 1.06$ μm, when mirrors manufactured of silver have practically no competitors; indeed, for all the other metals the quality parameters as computed with the aid of equation (4.17) are considerably lower at this wavelength. We note, however, that the use of Ag mirrors is not without problems. Let us mention here their rather weak resistance to thermoelastic damage, and in particular their rapid degradation, as an effect of the increase in $A(T_0)$, when in an environment containing sulphur compounds.

Table 4.3 Quality parameters, concerning the melting of some metallic mirrors exposed to CO_2 laser radiation.

Metal	F_G (10^3 J cm^2 s$^{-1/2}$)	$E_s^{m'}$			
		$\tau_p = 1$ ns	100 ns	1 μs	100 μs
Ag	642	18.1	181	574	5740
Al	154	4.31	43.1	136	1360
Au	489	15.7	157	431	4340
Cu	608	17.0	170	538	5380

4.5. Metal vaporisation

A moderate excess intensity over the I_m threshold is not generally accompanied by noticeable changes in the surface relief of mirrors. Indeed, for $I \geqslant I_m$ there exist no forces, with the exception of the pressure of light, directed normally to the surface and able to cause visible displacements of the liquid metal. As for the light pressure, we note that with CO_2 laser radiation of intensity $I = 10^8$ W cm^{-2}, its value is only $F = 0.03$ kg cm^{-2}.

However, due to the sharp modification in the optical and thermophysical properties of metals for $T > T_m$, a quite small amplification in the incident intensity over the I_m threshold may prove enough to cause a fast temperature excursion up to the boiling point, and the initiation of intense vaporisation of the target material (I_v denotes the threshold intensity characterising this process).

Let us comment, in a few lines, on the most important features of the vaporisation damage of metallic mirrors.

First, vaporisation damage causes the formation of a crater in the zone of action of the laser radiation. A typical photograph of such a crater is given in figure 4.8, along with a cross section of the irradiation zone. However, as in the case of melting, damage by vaporisation ($I = I_v$) can occur—depending on the material's purity and the degree of processing of its surface—either in point-like, localised zones, or simultaneously, over the whole irradiation spot.

Second, vaporisation of the metal proceeds from a liquid phase. A recoil, reactive pulse of vapours is therefore acting upon the melted layer, and the existence of a vapour gradient acting from the centre to the spot periphery leads to supplementary removal of metal from the irradiation spot [126, 127]. These effects provide an interpretation for the deposition of substance on the borders of the crater, well visible in figure 4.8.

Third, metal vaporisation is generally accompanied by plasma ignition in vapours, which facilitates the visual identification of the moment when vaporisation damage starts and eliminates the need for a thorough analysis of the sample's surface.

And finally, perhaps the most important phenomenon related to the mirror's resistance to light is the formation, within and around the irradiation zone, of a large variety of surface reliefs (see figure 4.8) whenever metal melting and/or vaporisation is induced. As a result, the optical quality of the mirror is catastrophically impaired, especially in the case of multipulse laser irradiation at $I \geqslant I_v$. Let us illustrate this, reporting the experimental data obtained in investigations of the dynamics of modification of the optical properties of a metallic mirror exposed to multipulse laser irradiation in vacuum.

Measurement of the specular reflectivity R_R during irradiation and after the laser is switched off is a convenient and appropriate method by which to control the optical resistance of the mirror. Indeed, the operational capacity

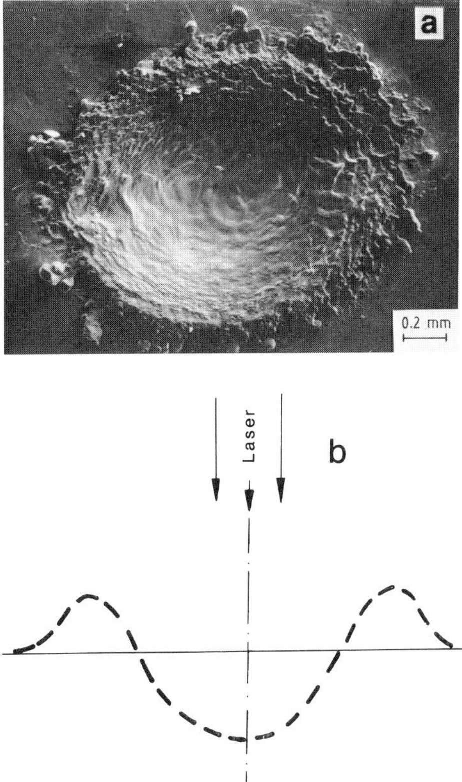

Figure 4.8 (*a*) Photograph of a crater formed in the irradiation zone on the surface of a gold sample. (*b*) A transverse section across the irradiation zone.

of a mirror must be considered impaired whenever its specular reflectivity R_R falls below its initial value R_R^0. The quantity R_R can be measured by collecting with a photodetector, or using a calorimeter, the fraction of the incident laser signal causing damage to the sample surface specularly reflected by the sample. Alternatively, one can record the reflection of another, probing, laser beam of lower power, which can be operated on a different wavelength, $\lambda_p \neq \lambda$.

In references [8, 135], a TEA CO_2 laser served as the main radiation source ($\lambda = 10.6\ \mu m$) generating pulses of two different durations: $\tau_p = 100$ ns or 3 μs. The samples exposed to irradiation were made of stainless steel or aluminium, and exhibited an initial roughness $\leqslant 0.3\ \mu m$. The reflectivity of the zone exposed to the irradiation from the main laser was probed with a second laser source emitting on a comparable wavelength $\lambda_p = 9.6\ \mu m$—implicitly assuming that the optical properties of the sample do not depend on wavelength within such a narrow range of wavelength variation. The intensity of the radiation from the main laser beam exceeded the vaporisation threshold of the samples ($I_0 > I_v$).

The dynamics of the relative specular reflectivity, R_R/R_R^0, during and after the action of a single laser pulse is presented in figure 4.9. As one can see from this figure, even monopulse irradiation would sometimes suffice to cause significant modifications in the specular reflectivity of mirrors. The lowest R_R values are reached at the end of the laser pulse action, and the drop in R_R has a non-reversible character (the R_R/R_R^0 curves in figure 4.9 are stationary by $t > \tau_p$).

A completely different situation is to be observed for intensities in the range $I_m \leqslant I_0 \leqslant I_v$. As one can see from figure 1.1 and 1.18(b), the temperature as well as any other modifications in the optical properties of the metal which are induced during laser irradiation show a reversible character. After the laser pulse one always finds $R_R \simeq R_R^0$.

An even larger decrease in R_R appears as an effect of multipulse irradiation of mirrors for $I \geqslant I_v$ [128]. This is visible in figure 4.10 and 4.11, where one

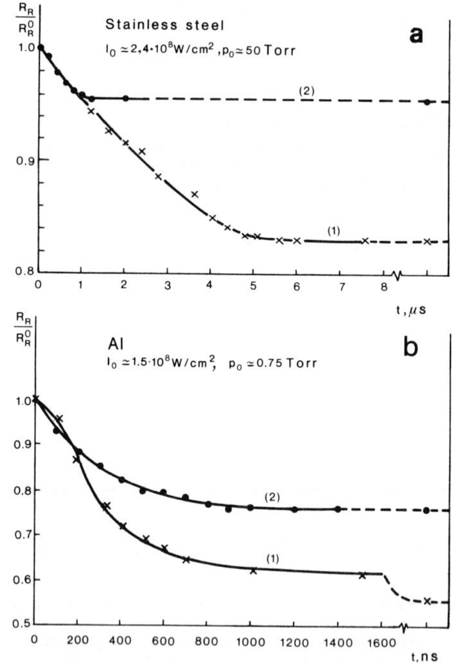

Figure 4.9 The evolution of the specular reflectivity in the light of a probing laser source ($\lambda_p = 9.6$ μm), for stainless steel (a) and aluminium (b) samples exposed to pulses generated by a TEA CO_2 laser source ($\lambda = 10.6$ μm), with durations $\tau_p = 3$ μs (curves 1) and $\tau_p = 100$ ns (curves 2), respectively.

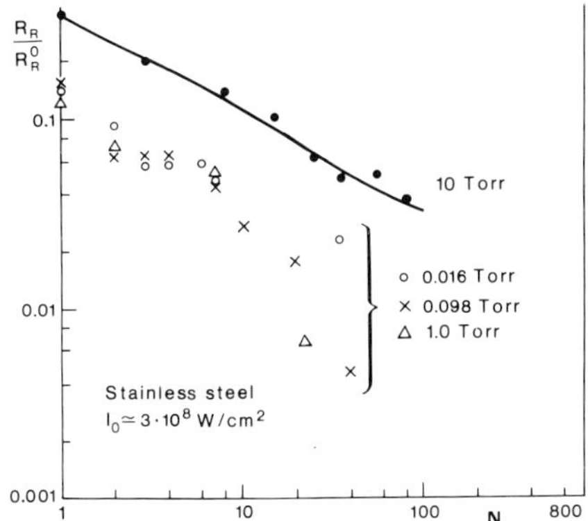

Figure 4.10 The dependence of the specular reflectivity of a stainless steel sample exposed to TEA CO_2 laser irradiation ($\lambda = 10.6\,\mu m$), on the number N of consecutive laser pulses directed on the same irradiation site, at different ambient air pressures.

represents the dependence of the relative specular reflectivity R_R/R_R^0 ($\lambda = 10.6\,\mu m$) on the number N of consecutive laser pulses ($\tau_p = 100$ ns) directed on the same irradiation site. Each pulse had a peak intensity at $I_0 \simeq 3 \times 10^8\,W\,cm^{-2}$. One can see that R_R can decrease by two orders of magnitude as compared with its initial value, for a sufficiently high number of consecutive laser pulses (at the given level of incident intensity, I_0). The dependence of R_R/R_R^0 on the pressure of the residual gas, visible in figure 4.10, is determined by the fact that, for $I_0 \simeq 3 \times 10^8\,W\,cm^{-2}$, and in the 1–10 Torr pressure range of ambient air, the optical discharge in the gas

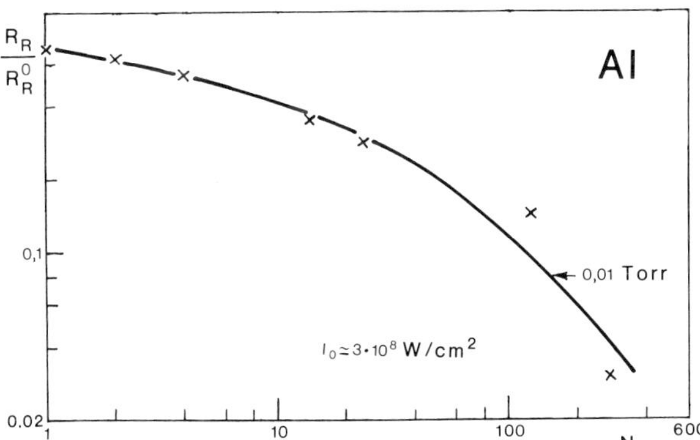

Figure 4.11 The dependence of the specular reflectivity of an aluminium sample exposed to TEA CO_2 laser irradiation in air ($p \simeq 0.01$ Torr), on the number N of subsequent laser pulses directed on the same irradiation site.

is induced close to the sample surface, while for lower pressures the plasma is initiated in metal vapours. In this second case, the R_R variation, as evidenced by measurements performed during the action of the laser pulses, is related not only to the modification from pulse to pulse of the surface relief, but also to the absorption and scattering by the plasma of a part of the pulse energy.

In conclusion, irreversible modifications of the surface relief are caused by laser irradiation at intensities $I_0 \geqslant I_v$, which are accompanied by a reduction in the specular reflectivity down to values depending on the number of consecutive laser pulses directed on the same irradiation site.

Let us note that irreversible damage of the surface is possible with multipulse laser irradiation at much lower levels of incident intensity, for $I_p^{\sigma_m} < I_0 < I_m$, due to cumulative effects [129]. We mention here the results reported in references [115, 130], where copper mirrors were exposed to consecutive laser pulses with a duration $\tau_p = 200$ ns and a repetition rate $f = 5/8$ Hz. It was shown that, for $I_0 > I_p^{\sigma_m}$ and long before catastrophic damage of the mirror or the appearance of a significant temperature increase on its surface, an increase in the roughness of the irradiation zone was induced,

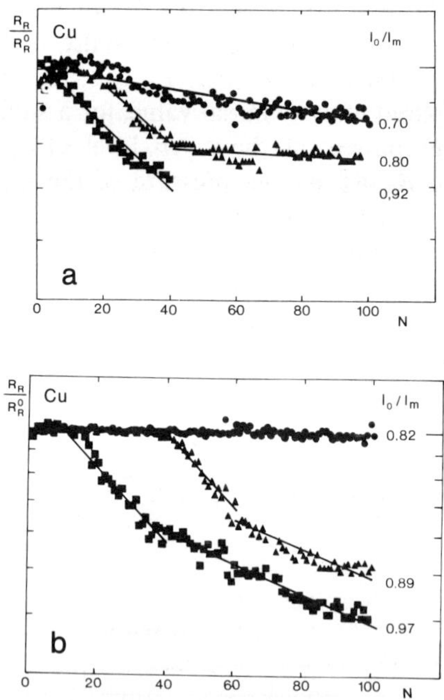

Figure 4.12 Reduction in the reflectivity of copper mirrors by multipulse laser irradiation below the melting threshold, $I_0/I_m < 1$. The duration of radiation pulses with $\lambda = 10.6\ \mu$m was $\tau_p \simeq 1.7$ ns. (*a*) Mirror with polished surface. (*b*) Mirror whose surface was processed by high-speed turning with a diamond tool.

from pulse to pulse. However, this modification of surface relief as a result of light-induced thermoelastic deformation of the metal causes a marked reduction in R_R after the action of a certain number N of consecutive laser pulses, only if a critical level of incident intensity is surpassed $I_p^{\sigma_m} < I_{cr} < I_m$—a value characteristic for a given mirror. In figure 4.12 the dependence on N of the ratio R_R/R_R^0 is represented, for different I_0/I_m values, and for two types of copper mirrors. One can clearly see a marked decrease in R_R appearing at intensities $I_p^{\sigma_m} < I_0 < I_m$, $I_0 \geqslant I_{cr}$. We also note that the reduction in I_0 means that a larger number of consecutive laser pulses are needed before irreversible thermoelastic damage takes hold of the metal surface.

4.6. Size effects

In problems dealing with the interaction of laser radiation with solid samples, by size effects one usually understands the dependence of the thresholds, or of the dynamics of certain processes, on the radius of the irradiation spot. There are many such size effects, expressed, for example, by the dependence on R_R of the temperature of the irradiated surface, for a given irradiation geometry and fixed sample dimensions.

In the following we shall consider several size effects occurring in the process of laser damage of the surfaces of metal mirrors.

The effects are obvious in the case of the quantities I_m and I_v, as these are obtained by solving the heat equation, $T(R_s, I_m) = T_m$ and $T(R_s, I_v) = T_v$ respectively. The functional relations $I_m(R_s)$, $I_v(R_s)$ therefore follow from the approximations used in solving the heat conduction equations.

An interesting size effect was reported in reference [115], where damage thresholds of metallic mirrors exposed to single-pulse or multipulse irradiation were investigated.

It was shown that the quantity $I_p \leqslant I_m$, where I_m is the melting threshold over the whole irradiation spot, depends in a completely different manner on R_s for the case of monopulse and multipulse action. The dependence $I_p(2R_s)$ for the two irradiation regimes is presented in figure 4.13.

The decrease in I_p with R_s, observed in reference [115] in the case of monopulse action, is due to the heating and subsequent destruction of surface defects. Accordingly, the probability of finding within the irradiation spot the required number of defects of a certain type and dimension increases with increase in R_s. This results in a reduction of I_p values (see, for example [138, 139]).

Alternatively, in the case of multipulse irradiation, if surface damage proceeds by a thermoelastic mechanism and the damage itself is determined by the σ_{rz} component of the stress tensor, then, according to equations (4.2) and (4.7), one has $I_p^{\sigma_m} \sim R_s$. This particular dependence $I_p(R_s)$ is in excellent

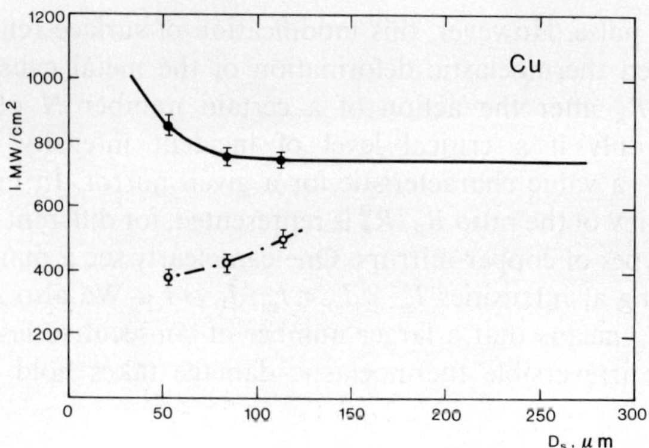

Figure 4.13 The dependence of the damage thresholds of a copper mirror on the diameter of the irradiation spot, for monopulse (full curve) and multipulse (chain curve) laser irradiation.

agreement with the experimental data (the chain curve in figure 4.13) observed with multipulse irradiation of samples, when one can take $f = 0$ in equation (4.7).

Chapter 5 Surface Periodic Structures

Two classes of laser-induced surface phenomena playing an important part in the laser heating of metals are discussed in the next three chapters, namely the surface structures and thermochemical processes in reactive environments. In this context Chapter 5 reviews experimental data and theoretical interpretations on laser-induced surface structures—also called laser ripples. New results are then introduced, on the laser beam interaction with self-induced or pre-existent surface structures. Enhancement of energy coupling, amplification of the electric field on the surface and wavelength-dependent size effects are shown to feature the interaction process, with interesting prospects for applications.

Many investigations concerned with surface damage by laser irradiation have indicated that the structures that form on surfaces can exhibit a considerable degree of order. One can divide such structures, observed on the surface of metals, dielectrics, semiconductors, into two categories.

(i) Resonant periodic structures (RPS), induced as a result of the interaction of the incident radiation with the electromagnetic waves that propagate across the surface (surface electromagnetic waves). Typically, RPS have periods determined by the wavelength (wherefrom their name), polarisation and incidence angle of the laser radiation.

(ii) Non-resonant periodic structures (NRPS). The characteristics of these cannot be directly related to the wavelength, nor to the degree and direction of polarisation of the laser radiation.

5.1. Non-resonant periodic structures (NRPS)

The observation of NRPS was most probably reported for the first time in reference [133]. As a result of the action of ruby laser pulses ($\lambda = 0.694\ \mu$m,

$\tau_p \simeq 10$ ns), NRPS with a period $\Lambda \sim 10\ \mu m$ were seen in the laser irradiation zone. More thorough investigations of NRPS have been reported in references [122–124, 134], using experiments performed with non-polarised radiation generated by pulsed CO_2 laser sources ($\tau_p = 1.7$ ns) [122–124] and by XeCl* excimer lasers ($\lambda = 0.308\ \mu m$; $\tau_p \simeq 30$ ns) [134]. The main results of these studies could be summed up as follows.

(i) For formation of NRPS, multipulse laser irradiation is required at the same site on the surface. Structures not exhibiting a clear degree of organisation (figure 5.1(*a*)) are usually observed within the irradiation spot, for small numbers, N, of consecutive laser pulses, while for larger N these can be identified only at the periphery of the irradiation spot. The self-organisation of the surface relief is increasingly evident as the number, N, of consecutive laser pulses increases, resulting in the formation of one-dimensional (figure 5.1(*b*)), or two-dimensional (figure 5.1(*c*)), surface periodic structures.

(ii) The period Λ of NRPS changes within the range 1–50 μm depending on radiation intensity, number, N, of consecutive laser pulses and target material. Moreover, structures with different periods $\Lambda \neq \lambda$ have been observed within the same irradiation spot.

(iii) Melting of the metal surface with every laser pulse is a fundamental requirement for NRPS formation, in both air and vacuum.

(iv) From the available experimental data it is evident that the transition from a disordered to an organised surface relief is always accompanied by a noticeable erosion of the sample material and by plasma ignition in vapours.

(v) The irradiation duration has to be short enough to 'freeze' the profile of the sample's surface formed in the liquid phase of the metal during its subsequent cooling.

(vi) The formation of NRPS is accompanied by a decrease in the specular reflectivity of the sample and leads to a lowering of the plasma ignition thresholds.

The mechanisms through which NRPS occur are not completely clear at this stage. One model that seems to be adequate is proposed in reference [134]. According to this model, when the surface relief is modified as a result of the non-uniform vaporisation of the metal, self-organisation of the relief takes place due to the instabilities evolving at the plasma–target interface (plasma interacts with spontaneous intense electric fields in conditions of capillary oscillations of the molten surface layer).

Concluding this section, we mention that, as was shown in reference [127], NRPS may also occur due to the interaction of polarised light with surfaces. In this case, RPS develop and persist in the irradiation spot up to a certain critical number of consecutive laser pulses, $N = N_{cr}$, when they begin to alter their nature, gradually becoming NRPS.

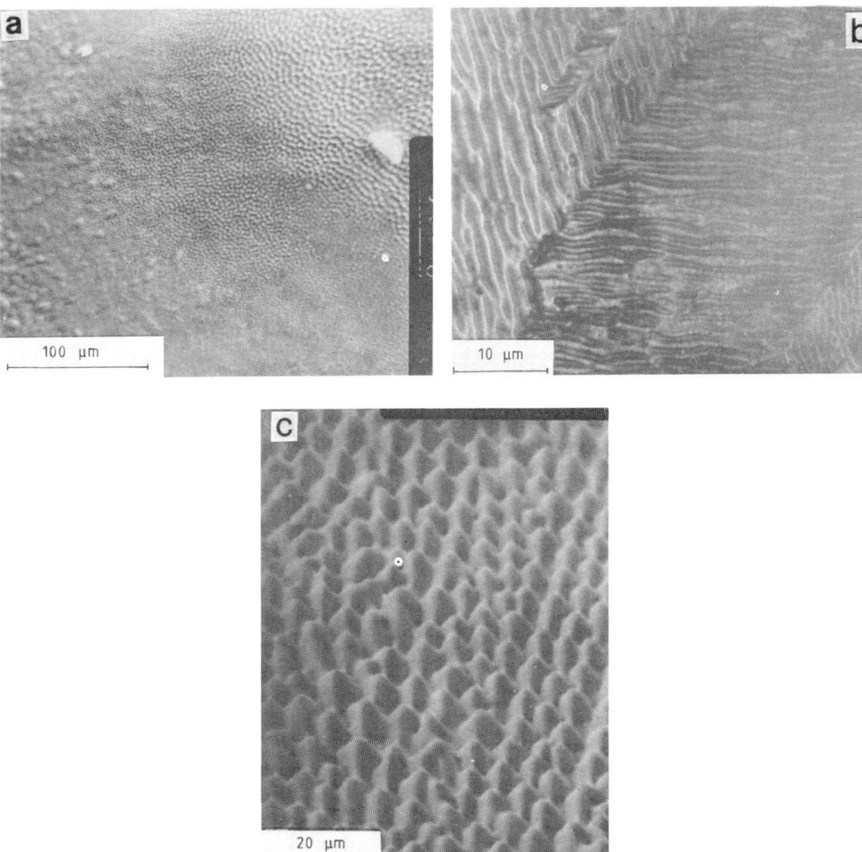

Figure 5.1 Non-resonant periodic structures (NRPS) induced on surfaces of metallic samples as a result of multipulse XeCl* excimer laser irradiation ($\lambda = 0.308$ μm), in air. (*a*) Disordered NRPS at the border of the irradiation spot on the surface of electrolytic copper (peak pulsed intensity $I_0 = 200$ MW cm^{-2}; the number of consecutive laser pulses, $N = 200$). (*b*) One-dimensional NRPS on the surface of a brass sample ($I_0 = 120$ MW cm^{-2}, $N = 100$). (*c*) Two-dimensional NRPS on the surface of a lead sample ($I_0 = 30$ MW cm^{-2}, $N = 100$).

5.2. Resonant periodic structures (RPS)

Such structures were observed in the very first microscopical investigations of the surfaces of samples that underwent melting under the action of intense, linearly polarised, laser radiation [135–147]. In the case of a uniformly smooth surface, and for a uniform distribution of energy across the irradiation spot, RPS appeared in the form of a band pattern—a sort of linear diffraction grating—with a period, at perpendicular incidence of the radiation, $\Lambda \simeq \lambda$. The grooves and prominences thus formed are oriented perpendicular to the electric field of the light wave. It has been noted that RPS can result under the action of either single-pulse or multipulse laser irradiation. The generation of RPS proceeds from various types of surface defects: cracks, grain borders, etc. An example of RPS obtained by single-pulse laser irradiation of a metal sample is given in figure 5.2(*a*). One can see that close to a surface defect the relief froze in the form of a spherical wave—a typical example of NRPS. However, far from the centre of this wave the pattern evolved into a

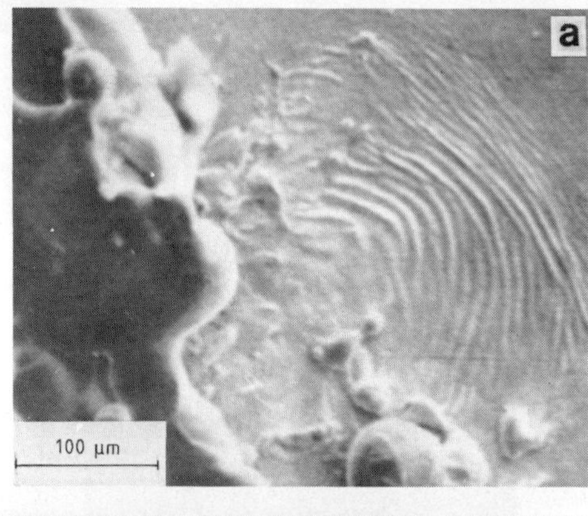

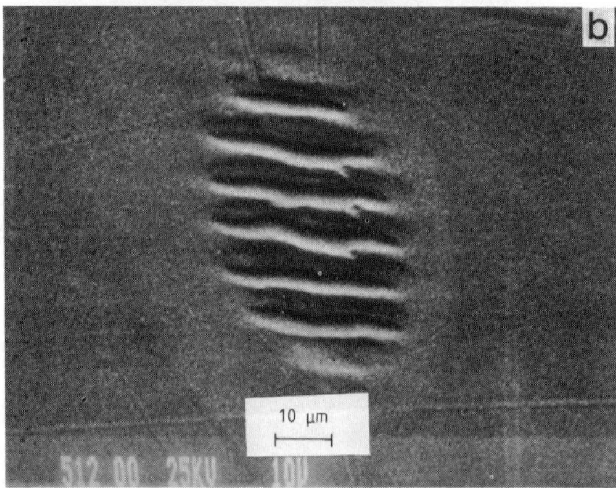

Figure 5.2 Resonant periodic structures (RPS). (*a*) Formation of RPS in the vicinity of a surface defect, by surface melting of a copper sample under the action of single-pulse TEA CO_2 laser radiation ($I_0 \simeq 5 \times 10^8$ W cm^{-2}). (*b*) Well developed RPS, generated by multipulse TEA CO_2 laser irradiation of a molybdenum sample in vacuum (the peak radiation intensity for all laser pulses was $I_0 \simeq 5 \times 10^7$ W cm^{-2}, and the number of consecutive laser pulses was $N = 200$).

linear configuration, with a period close to the radiation's wavelength, $\Lambda \sim \lambda$, which indicates an RPS.

Along with the increase in the number of radiation pulses directed at the same area on the surface, the relief formed on the surface sharpens, and also undergoes transformations: the spherical waves in the neighbourhood of point-like defects, or those oriented, for example, parallel to cracks, are gradually wiped out. As a result, what persists on the surface is a linear pattern oriented perpendicular to the electric field of the light wave (figure 5.2(*b*)).

In the case of oblique incidence of linearly polarised radiation, RPS are modified. If the radiation is polarised in the incidence plane, we have

$$\Lambda = \Lambda_{\parallel} = \frac{\lambda}{1 \pm \sin\theta}. \tag{5.1}$$

In the case of radiation polarised perpendicular to the incidence plane

$$\Lambda = \Lambda_{\perp} = \frac{\lambda}{\cos\theta}. \tag{5.2}$$

Significant progress in the investigation of RPS structures was made possible by the analysis of the diffraction pattern obtained when illuminating the irradiation zone with a probing laser source [143–154]. The diffraction pattern, corresponding to the surface observation in the wavevector space forming by diffraction on RPS gratings, was found in many cases to provide more information than the image of the irradiation zone in the real space. Such investigations have shown that the range of periods as well as the range of directions are far wider than those indicated by equations (5.1) and (5.2), though the structures with periods $\Lambda(\theta)$, which can be determined from equations (5.1) and (5.2), are still predominant at small angles of incidence. It was also determined [155], that RPS can be obtained not only under the action of linearly polarised radiation, but also with circularly polarised radiation. However, growth (deepening into the sample) of the structures under the action of circularly polarised radiation proceeds at a much lower rate than in the case of linearly polarised radiation.

Figure 5.3 [150, 151] presents schematic diffraction patterns (reflectograms) obtained in several typical situations. For $\theta \neq 0$ these consist of two intersecting circles of radius $2\pi/\lambda$ (in the wavevector space) placed at a distance of $(2\pi/\lambda)\sin\theta$ from each other. From the physical point of view, the grating of vector $\boldsymbol{g}(|\boldsymbol{g}| = 2\pi/\lambda)$ is equivalent to a grating of vector $-\boldsymbol{g}$. That is why the diffraction image is always symmetrical as against its centre. As one can see from figure 5.3, the diffraction patterns do not consist of discrete points corresponding to the above-mentioned RPS, but rather of circular arcs, i.e. they contain not only gratings with grooves oriented perpendicularly to the vector of the incident laser field, $\boldsymbol{E}_i$ (when we have $\boldsymbol{g} \parallel \boldsymbol{E}_i$), but also gratings with $\boldsymbol{g}$ vectors which are not parallel to $\boldsymbol{E}_i$. We point

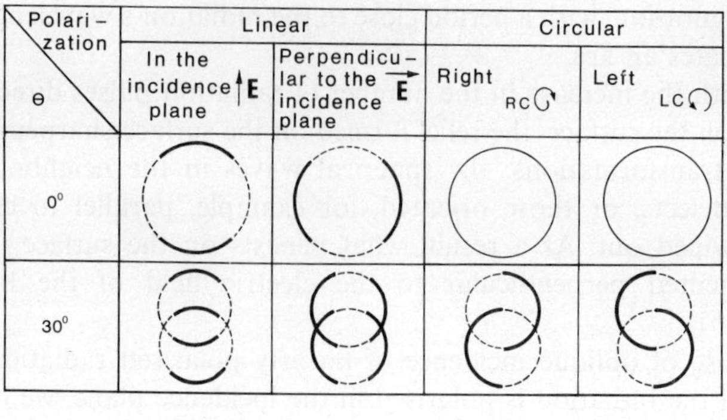

Polari–zation / θ	Linear		Circular	
	In the incidence plane E	Perpendicu–lar to the incidence plane E	Right RC	Left LC
0°				
30°				

Figure 5.3 Diffraction patterns (reflectograms) of RPS for different polarisation states and incidence angles θ of laser radiation on a sample.

out, however, that such reflectograms were obtained on well established RPS formed after multipulse laser irradiation when their structure no longer depends on the initial conditions.

It is interesting to note that under the action of circularly polarised laser radiation incident perpendicularly on the sample surface ($\theta = 0°$), structures are forming with a $\Lambda = \lambda$ period in all directions.

For the above-described diffraction patterns one has $|\boldsymbol{g}| < 2|\boldsymbol{k}_0|$, where $|\boldsymbol{k}_0| = 2\pi/\lambda$. However, in a series of experiments on germanium at $\lambda = 10.6\ \mu m$ [145, 156] RPS were observed with $|\boldsymbol{g}| = (4\text{–}6)|\boldsymbol{k}_0|$, i.e. with shorter periods, $\Lambda \simeq \lambda/(4\text{–}6)$. In this case $\boldsymbol{g}$ is perpendicular to $\boldsymbol{E}_i$.

The reflectogram method enabled the monitoring of RPS kinetics during the action of the laser pulse [157, 158]. It was shown that besides the 'frozen' structures that persist after irradiation, there also exists a reversible grating that lasts only as long as the interaction process itself. Such reversible gratings were observed on liquid metals [158] and molten semiconductors [157] as well as on initially hard surfaces. The space and time evolutions of RPS from pulse to pulse was monitored by reflectograms, as well as by microscope viewing [159].

An important result of these experiments was also the discovery that the development of RPS—at least during the early stages of their formation—follows the pattern of an exponential amplification in time of instabilities in the surface relief [149]. The dynamics of their development is analogous to that known for the non-linear instabilities occurring in the case of Raman scattering.

5.3. Mechanisms of RPS formation

For a theoretical analysis of the nature of the phenomena related to the laser-induced RPS, the problem can be approached by two interconnected stages:

(i) electrodynamic analysis, which must make clear the nature of the wave propagating along the sample's surface, and which, by interference with the incident wave, modulates the energy absorbed into the metal;

(ii) thermophysical analysis, investigating the very process that determines the modification of the surface relief as an effect of the non-uniform energy dissipation.

Most of the work on RPS has tackled only the first stage of this approach.

As a wave which, by interference with the incident wave, causes an intensity modulation, many authors have proposed a surface-scattered wave with a $|\boldsymbol{k}_0| = 2\pi/\lambda$ wavevector, initiated on surface defects and/or cracks. Although this relatively simple concept served well the prediction of some experimentally confirmed results, it was eventually criticised [151, 160] on the grounds that such a surface-scattered wave would be unable to satisfy Maxwell's equations.

Another type of wave which could cause a spatial modulation in the intensity are the surface electromagnetic waves (SEW) [148, 156]. This model can be applied only in the case of media that are active in relation to the excitation of polaritons and plasmons, when the following condition is fulfilled

$$\mathrm{Re}\ \varepsilon_c < -1 \tag{5.3}$$

where ε_c is the complex dielectric permittivity of the material for the respective incident radiation wavelength. Condition (5.3) is fulfilled for all wavelengths of power lasers in current use, in the case of metals and several semiconductors, for $T > T_m$. However, SEW cannot explain RPS formation in dielectrics such as NaCl, KCl, BaF_2, SiO_2, etc [144, 146, 148] or in the case of semiconductors subjected to limited laser heating [161].

The generation of a scattered wave is also possible owing to the existence on the surface of the target of a thin, layered, waveguide. For example, such a waveguide, 50 μm in thickness, was obtained, according to the authors of reference [161], on the surface of a GaAs sample by etching with a $H_2SO_4/H_2O_2/H_2O$ solution, and under the action of an He–Cd laser source. According to reference [156], a waveguide is also present on the surface layer of a germanium sample exposed to the action of CO_2 laser pulses, due to the temperature gradient induced by pulsed laser irradiation. This phenomenon offered a way to explain the formation of periodic structures with $\boldsymbol{g}$ perpendicular to $\boldsymbol{E}_i$ and a period $\Lambda \simeq \lambda/n$, where $n = 4$ is the refractive index of germanium. Another example of waveguide structure is a metallic sample covered by an oxide layer.

The authors of references [150, 151, 152, 154, 160] have proposed a more general approach to the problem of the scattered wave. It was determined that the observed damage is the result of the action of electromagnetic fields created on surface roughnesses. However, due to the interface between the vacuum and the bulk material of the target, these fields are of the non-radiative kind. The Fourier component of the roughness with a vector $\boldsymbol{g}$ leads to the appearance of a Fourier component of the dipole moment (per unit area of surface layer) having a wavevector resulting from the relations

$$\boldsymbol{k} = \boldsymbol{k}_t + \boldsymbol{g} \qquad \boldsymbol{k}_t = \boldsymbol{k}_0 \sin\theta. \tag{5.4}$$

The field generated by such a dipolar layer can interfere beneath the irradiated surface with the refracted portion of the incident beam, causing a non-uniform absorption of energy. In this case the period of the grating roughness coincides with the period of the intensity grating created by it. The detailed computations show that the components of $\boldsymbol{g}$ for which

$$|\boldsymbol{k}_t \pm \boldsymbol{g}| = 2\pi/\lambda \tag{5.5}$$

are the most important for surface damage.

In reference [162] an approach was suggested to ascertain the mechanism that leads to the establishment of the positive feedback required for RPS to grow out of the random noise of space harmonics of the surface roughness. Based upon the examination of the model situation of a grating with $\Lambda = \lambda/(1 \pm \sin\theta)$, it was shown that the grating which directs the light waves diffracted in $+1$ or -1 orders into the material, creates in turn, by interference of these waves with the refracted beam, an intensity grating of the same period Λ, yet shifted by a phase Φ as compared with the initial grating (see figure 5.4). The quantity Φ is a function of Λ; the resonance value, $\Lambda_0 = \lambda/(1 \pm \sin\theta)$, undergoes important changes because of the change in Λ.

The thermophysical mechanism leading to RPS formation (thermal expansion, melting, vaporisation, etc) and the conditions for feedback to appear, ensuring a sharpening of the grating, determine the choice of the value of Φ, or the value of Λ in the vicinity of Λ_0, in the form of the dependence

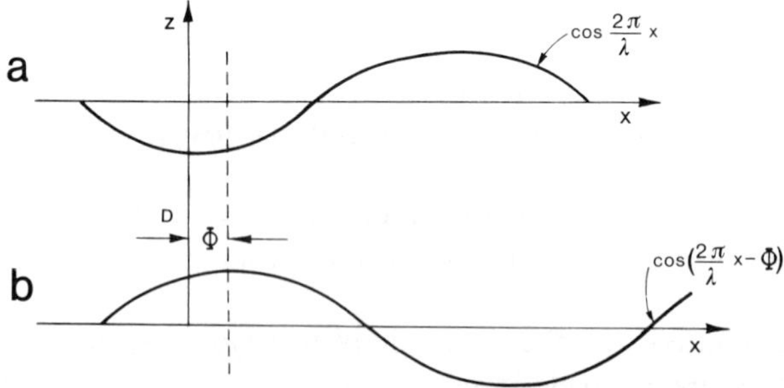

Figure 5.4 (*a*) The profile of a light-induced RPS. (*b*) The spatial profile (along the x axis contained in the sample surface, and directed perpendicular to the grating grooves) of the modulation of the radiation intensity.

$\Phi(\Lambda)$. For example [162], the grating profile can be described in the form

$$z(x) = -h\cos\frac{2\pi}{\Lambda}x \tag{5.6}$$

(here h indicates the relief depth) and that of the intensity grating in the form

$$I(x) = I_0\left[1 + I_1\cos\left(\frac{2\pi}{\Lambda}x - \Phi\right)\right]. \tag{5.7}$$

In the case of RPS formation due to the temperature dependence of the surface tension coefficient, it is necessary that RPS depressions be more strongly heated than their peaks in order to provide for the positive feedback. Then the liquid will be evacuated from the RPS depressions, where the superficial tension coefficient is lower than on the peaks, while RPS are even further sharpened. Correspondingly, for a grating grown in this way it is necessary that

$$-\frac{\pi}{2} < \Phi < \frac{\pi}{2}. \tag{5.8}$$

In contrast with the above, in the case of RPS formation by thermal expansion, in order to ensure the positive character of the feedback process it is necessary that RPS peaks be more strongly heated than the depressions and, therefore,

$$\frac{\pi}{2} < \Phi < \frac{3\pi}{2}. \tag{5.9}$$

From the examples given one can see that the electrodynamical and thermophysical parts of the problem are actually related to each other. The range of variation of electrodynamically determined values of Φ may limit the number of physically possible thermophysical mechanisms to provide for the sharpening of the relief of the irradiated surface.

In references [163–165] it was shown that due to the phase shift, Φ, as the RPS advances more deeply it can also move from pulse to pulse across the surface. We also note the attempt to explain by an electrodynamical approach [166–168], a series of surface effects of the optical constants n, k; in particular, areas were identified where there are RPS with $\boldsymbol{g}$ parallel to $\boldsymbol{E}_i$ or $\boldsymbol{g}$ perpendicular to $\boldsymbol{E}_i$. However, the examination of the problem in the framework of an electrodynamical approach alone is not always correct. In reference [169] it was thus shown that whenever $|\varepsilon_c| \sim 1$, the calculation of period can be performed only by completely solving the problem. Attempts were also recorded to interpret the generation of some types of RPS by means of the scattering of radiation by the spatial modulation of the dielectric permittivity [162, 170], and not by the surface roughness. Also worthy of mention are references [171, 172], where the formation of RPS was explained on the basis of acoustic waves induced by the frequency beats of the longitudinal modes of the laser radiation. In the case of GaAs irradiated with ruby laser pulses one obtains $\Lambda \simeq \lambda/2$.

A simultaneous, correlated examination of both the electrodynamical and thermophysical parts of this problem was performed in references [145, 168, 169, 173–176]. The main stages in the computation of RPS induction can, in this case, be described as follows.

The diffraction of the $\boldsymbol{E}_i$ incident laser field by the space–time Fourier component of the surface relief of the $\boldsymbol{g}$ wavevector is solved in the first, electrodynamical, stage of the analysis. In an approximation that retains only contributions linear in the amplitude ξ_g, the expressions for the diffraction fields inside the substance are determined. The Q_g amplitude of the spatially non-uniform (across the surface) distribution of the power density of the heat source, appearing as a result of interference of the incident and diffracted waves, is then determined.

In the second, thermophysical, stage one determines the spatial distribution of temperature—which further causes the induction of forces modifying ξ_g. In this way, by solving one or other equation relating ξ_g to the temperature, one obtains the feedback chain causing RPS development. When discussing surface acoustic waves the appropriate equation is the equation for the vector of elastic deformation of the medium; for capillary waves on molten surfaces one is bound to use the hydrodynamic equations [177]; in case of interferential instability of the vaporisation front one should use the equation of displacement of the vaporisation front with solid–vapour or liquid–vapour borders [173, 177, 178].

We also mention some other mechanisms [179] of relief modification, for example chemical ones, such as oxidation, etching and deposition from the gaseous phase.

The use of the feedback in the examination of RPS formation allows for the calculation of the increment in time, γ_g, of the exponential increase (or damping) of the relief amplitude. The dependence on $\boldsymbol{g}$ of the γ_g maxima and the $\boldsymbol{g} = \boldsymbol{g}_{max}$ values determine the periods $\Lambda = 2\pi/|\boldsymbol{g}_{max}|$ and the directions of the dominating grooves of RPS. In contrast with the general case, solving only the electrodynamical part of the problem does not allow us to find the dominating type of RPS, but would rather indicate the possible types.

Ending this section we emphasise that in the case of interest, namely RPS generation on metals, the main thermophysical mechanisms causing the growth of surface relief, are:

(i) thermocapillary instability and/or metal vaporisation—of course for these to occur the metal must melt while for the observation of the structures at the cessation of the laser pulse a fast quenching of the sample is required to freeze the relief that was formed during irradiation;

(ii) the chemical processes on the irradiated surface in the field of the space-modulated radiation.

5.4. The absorptivity of a rippled surface

Light-induced RPS have created considerable excitement. One interesting feature related to RPS is the active influence that they can exert upon the optical properties of the surface on which they grow. Thus it will be shown later that, as the surface relief on the metal surface sharpens, target absorptivity for the incident radiation can markedly increase, and even reach 100%. Though there is no direct experimental evidence available as yet of such a radical change in the optical characteristics of surfaces, numerous experimental data on 'abnormally' high absorptivity of metals by $I \geqslant I_m$ (I_m being the threshold for surface melting) would sustain such a possibility (see figure 1.18).

For the sake of simplicity we shall first examine the static problem, i.e. the case when the area of the irradiation spot is covered by pre-existing RPS of given and invariable parameters. We shall discuss the simplest case, when an analytical solution is possible [180, 181].

Let us consider a light beam of parallel rays (figure 5.5), with a rectangular cross section, linearly polarised in the zx incidence plane and falling under an incidence angle θ, measured with respect to the perpendicular on the surface of the metallic sample. The irradiation spot of area $S_s = x_0 y_0$ is covered with a diffraction grating having a sinusoidal profile $z = h \sin gx$, where h is the profile amplitude, a quantity independent of the x coordinate. For the chosen orientation of the grating grooves (figure 5.5) and a Λ value determined from equation (5.1), the diffraction grating behaves resonantly with respect to surface electromagnetic wave (SEW) excitation. We shall take in the following $\Lambda = \lambda/(1 - \sin\theta)$, although analogous results are obtained also in

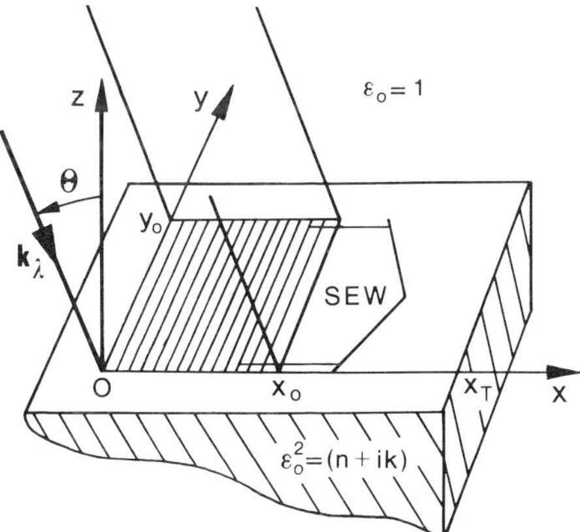

Figure 5.5 The interaction scheme, preparing the ground for the model used to calculate the optical properties of a sample with a resonant periodic structure inside the irradiation spot.

the case $\Lambda = \lambda/(1 + \sin\theta)$. The parameters of the problem do not depend on the y coordinate as the energy distribution within the irradiation spot is uniform. And finally, the sample surface is plane outside the irradiation spot and the sample dimension along the $0x$ axis (the SEW propagation direction) is $x_T \geqslant x_0$.

Then, according to reference [174], SEW excitation is described by the equation

$$\frac{da}{dx} = \beta f - \alpha a \tag{5.10}$$

where f is the amplitude of the electrical field of the incident wave, $a(x)$ is the amplitude of the H_y magnetic field of SEW for $z = 0$, β is a coefficient linking the incident wave and SEW on the grating ($\beta \sim kh$), while the attentuation coefficient of SEW is to be obtained from the expression

$$\alpha = \alpha_d + \alpha_r \tag{5.11}$$

where α_d is determined by the SEW energy dissipation as heat, and α_r takes into account the radiative losses (on the plane surface, i.e. in our case outside the irradiation spot, we have $\alpha_r = 0$).

In the general case, the solution of equation (5.10) takes the form

$$a(x, y) = \exp(-\alpha x) \int_{-\infty}^{\infty} \exp(\alpha\xi) f(\xi, y) \beta(\xi)\, d\xi. \tag{5.12}$$

Then by $f = \text{const}(x \leqslant x_0)$ and $f = 0(x > x_0)$ and for $\beta(\xi) = \beta = \text{const}$, one obtains from (5.12)

$$a(x, y) = \begin{cases} \dfrac{\beta}{\alpha} f[1 - \exp(-\alpha x_0)] & \text{for } 0 \leqslant x \leqslant x_0 \\[2ex] \dfrac{\beta}{\alpha} f[1 - \exp(-\alpha x_0)] \exp[-\alpha_d(x - x_0)] & \text{for } x_0 \leqslant x \leqslant x_T. \end{cases} \tag{5.13}$$

Function $a(x)$ is represented in figure 5.6. We notice that its maximum is reached at the border of the irradiation spot, and also that SEW amplitude does not vanish even in regions far from the irradiation spot. It means that, when SEW are present, laser energy dissipation into the sample proceeds not only inside the irradiation spot (as an effect of the absorption of incident radiation coinciding with heat dissipation by SEW), but also outside it as a result of SEW dissipation.

The power carried by SEW across a given $x = \xi$ surface results, in the case examined, from

$$P_{SEW}(\xi) = \frac{c_0}{8\pi} \delta_{SEW}\, y |a(\xi)|^2 \tag{5.14}$$

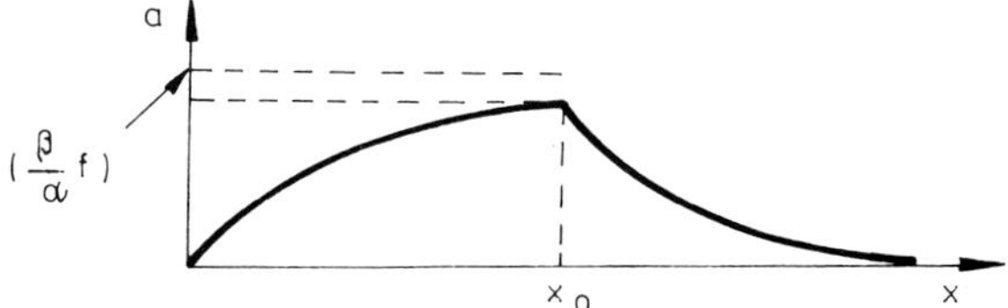

Figure 5.6 The evolution of the amplitude of a SEW field, excited on resonant periodic structures inside the irradiation spot of dimension x_0 (x is the coordinate along the sample surface).

where

$$\delta_{\mathrm{SEW}} = \frac{1}{2k\,\mathrm{Im}(\varepsilon_c + 1)^{-1/2}} \tag{5.15}$$

is the SEW attenuation depth along the z axis outside the metal in a medium of dielectric permittivity $\varepsilon_0 = 1$; ε_c is the complex dielectric permittivity of the metal, $\sqrt{\varepsilon_c} = n + ik$.

Taking into account that the power of the incident radiation can be obtained from

$$P_0 = \frac{c_0}{8\pi} f^2 y_0 x_0 \cos\theta \tag{5.16}$$

and using equations (5.13)–(5.16), one can calculate the contribution of the SEW energy dissipation to the overall absorptivity value

$$A_{\mathrm{SEW}}(x_0, x_T) = \frac{Q(x_0, x_T)}{P} = \frac{Q_1(x_0)}{P} + \frac{Q_2(x_0, x_T)}{P} \tag{5.17}$$

where

$$\begin{aligned} Q(x_0, x_T) &= Q_1(x_0) + Q_2(x_0, x_T) \\ &= 2\alpha_d \int_0^{x_0} P_{\mathrm{SEW}}(\xi)\,d\xi + 2\alpha_d \int_{x_0}^{x_T} P_{\mathrm{SEW}}(\xi)\,d\xi \end{aligned} \tag{5.18}$$

is the SEW power loss inside (Q_1) and outside (Q_2) the irradiation spot.

Equation (5.18) allows for the calculation of the total absorptivity of the sample

$$A \simeq A_0 + A_{\mathrm{SEW}} = A_0 + A_1 + A_2 \tag{5.19}$$

where A_0 is the absorptivity of the plane metal surface, while $A_1 = Q_1/P_0$ and $A_2 = Q_2/P_0$ are the efficiencies of energy conversion of SEW energy into heat inside and outside the irradiation spot, respectively. From equations

(5.12)–(5.18) we infer the expressions for A_1 and A_2

$$A_1(x_0) = F_1(\alpha x_0)G(h) \tag{5.20}$$

$$A_2(x_0, x_T) = \frac{\alpha}{\alpha_d} F_2(\alpha x_0)G(h)\{1 - \exp[-2\alpha_d(x_T - x_0)]\} \tag{5.21}$$

where functions F_1, F_2 and G are to be determined from the following simple expressions

$$F_1(\alpha x_0) = [2\alpha x_0 - 3 + 4\exp(-\alpha x_0) - \exp(-2\alpha x_0)]/(2\alpha x_0) \tag{5.22}$$

$$F_2(\alpha x_0) = [1 - \exp(-\alpha x_0)]^2/(2\alpha x_0) \tag{5.23}$$

$$G(h) = M\frac{4h_0^2h^2}{(h_0^2 + h^2)^2} \qquad M \simeq 1. \tag{5.24}$$

Functions $F_1(\alpha x_0)$ and $F_2(\alpha x_0)$ are represented in figure 5.7, while function $G(h)$ is given in figure 5.8. We observe differing evolutions of F_1 and F_2 with the increase in αx_0: while function $F_1(\alpha x_0)$ increases monotonically with the increase in αx_0 and tends to a maximum $\simeq 1$, function $F_2(\alpha x_0)$ reaches a peak for $\alpha x_0 \simeq 1.26$, and then decreases to zero. Quantity $G(h)$ shows a maximum, $G(h = h_0) = 1$, at a value of the grating depth

$$h = h_0 = \frac{\Lambda}{2\pi}\sqrt{A_0 \cos\theta}. \tag{5.25}$$

And finally we give the expression for

$$\alpha = \alpha_d\left(\frac{h_0^2 + h^2}{h_0^2}\right). \tag{5.26}$$

Let us further discuss equations (5.20), (5.21) in two limiting situations: $\alpha x_0 \gg 1$ (large irradiation spots) and $\alpha x_0 \ll 1$ (small irradiation spots).

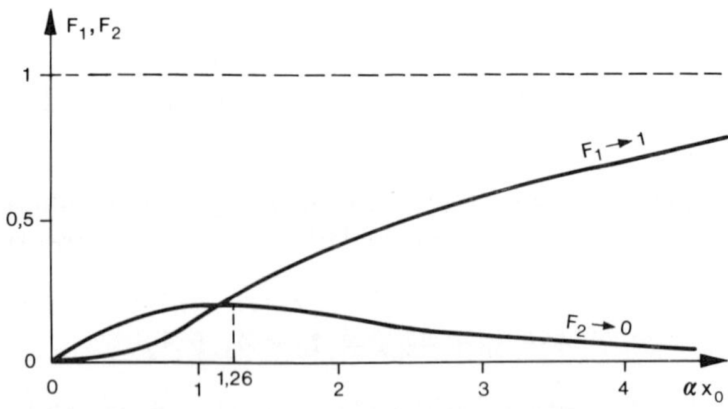

Figure 5.7 Dependence on αx_0 of the dimensionless functions F_1 and F_2.

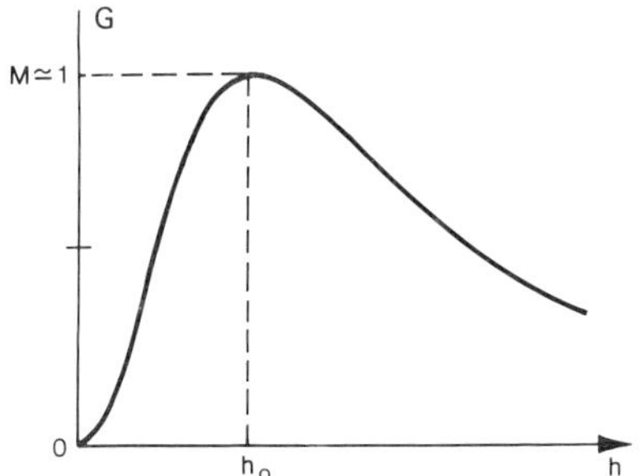

Figure 5.8 Dependence on the grating depth, h, of the dimensionless function G.

To this purpose, the values of interest of the various physical parameters are given in table 5.1.

The data in table 5.1 give, first of all, proper indications as to what x_0 values would qualify an irradiation spot as 'large'. We notice that with CO_2 laser radiation and for metals exhibiting a high electrical conductivity (Al, Ag, Cu), we have $\alpha_d \simeq 0.1\ \text{cm}^{-1}$. Then, assuming $h = h_0$, we get $\alpha \simeq 2\alpha_d \simeq 0.2\ \text{cm}^{-1}$, and for the fulfilment of the condition $\alpha x_0 \gg 1$ the dimension of the irradiation spot is $x_0 \geqslant 5$ cm. For metals of lower electrical conductivity, and especially for shorter radiation wavelengths, one notices an increase in the coefficient α_d, which for the radiation of $\lambda \sim 1\ \mu$m

Table 5.1 Values of some physical parameters of interest for the characterisation of the radiation absorption by SPS rippled surfaces of different metals.

Metal	λ (μm)	T (K)	n	k	$A_0 \times 10^{-2}$	α_d (cm^{-1})	h_0 (μm)	h^* (Å)
Al	10.0	293	31.2	104	1.06	0.15	0.164	84
	10.6	933 + 0			4.34	0.93	0.35	366
	0.95	293	1.75	8.5	9.3			
Cu	10	293	11.6	60.3	1.23	0.31	0.177	98
	0.95	293	0.13	6.22	1.34	35.5	0.0175	10
Ag	10	293	5.21	71.9	0.4	0.087	0.1	32
	0.95	293	0.11	65.6	1.02	25.6	0.015	1.7
Fe	10	292	1.00	25.6	3.98	2.27	0.32	317
	1	293	2.3	4.52	35.8			
Ti	10	293	1.85	18.5	1.77	5.59	0.444	618
	1.09	293	3.5	4.02	49.3			
W	10	293	8.25	41.5	1.84	0.67	0.22	146
Pb	11	293	23.2	39.2	4.47	1.2	0.37	390
	1	293	1.38	5.32	18.3			

wavelength may reach ~ 100–$1000\ \mathrm{cm}^{-1}$; consequently, the irradiation spot can be considered large for dimensions of only hundreds or even tens of micrometres.

Whenever the condition $\alpha x_0 \gg 1$ is fulfilled, one has $F_1 \to 1$ and $F_2 \to 0$, and from equations (5.20) and (5.21) one gets

$$A_1 \simeq G(h) \qquad A_2 \simeq 0 \qquad \text{for } \alpha x_0 \gg 1. \tag{5.27}$$

The maximum $A_1 \simeq 1$ is reached for a relief depth $h = h_0$ given by the expression (5.25). The computed h_0 values for several metals are given in table 5.1. We emphasise that $h_0 \ll \lambda$, and for metals such as Al, Ag and Cu, we have $h_0 \approx 0.1\ \mu\mathrm{m}$ by $\lambda = 10\ \mu\mathrm{m}$. Moreover, the further deepening of the relief ($h > h_0$) is accompanied by a decrease in A_1. Consequently, in the case of large irradiation spots, one can achieve experimentally the conditions for all laser energy to be practically transformed into SEW, and then be absorbed by the rippled surface.

We can introduce one more characteristic relief depth h^*, proceeding from the condition that the supplementary energy dissipation into the sample, caused by SEW generation, be equal to A_0:

$$A_1 = A_0(\theta) = \frac{A_0(\theta = 0)}{\cos\theta}. \tag{5.28}$$

Then, assuming that $(h^*)^2 \ll h^2$, we obtain

$$h^* = \frac{A_0(\theta = 0)}{4\pi}\Lambda. \tag{5.29}$$

Several computed values of h^* are given in table 5.1. Taking into account that (in case of CO_2 laser radiation) for most metals we have $A_0(\theta = 0) \sim 10^{-2}$, one obtains directly from equation (5.29) that $h^* \sim 10^{-3}\Lambda \sim 100$ Å. Also, as the radiation wavelength decreases to $\lambda \sim 1\ \mu\mathrm{m}$, the h^* value for metals Al, Ag and Cu is even lower, $h^* \sim 10$ Å, i.e. it corresponds to the surface roughness of the best present-day metallic mirrors. Therefore, the existence of such mirrors for which the spatial harmonics of the roughness are resonant to the incident radiation must lead to a significant increase in the metal mirror absorptivity, to $A = A_0 + A_{\mathrm{SEW}}$.

The condition $\alpha x_0 \ll 1$ results in a clearly expressed size effect. Indeed A_2 differs from zero, and while A_1 still depends only on x_0, A_2 depends not only on x_0 but also on the target dimension x_{T} (according to equations (5.22) and (5.23), respectively).

The infinite sample provides the simplest case for the discussion of small irradiation spots. We have then

$$2\alpha_{\mathrm{d}}(x_{\mathrm{T}} - x_0) \gg 1 \tag{5.30}$$

i.e. target dimensions would greatly exceed the dimensions of the irradiation spot. Then the expressions for A_1 and A_2 take the following simple forms

$$A_1 = \frac{(\alpha x_0)^2}{3} G(h) \tag{5.31}$$

$$A_2 = \left(\frac{\alpha}{\alpha_d}\right) \frac{\alpha x_0}{2} G(h). \tag{5.32}$$

By taking into account that $\alpha x_0 \ll 1$ and $\alpha/\alpha_d = 1 + h^2/h_0^2 > 1$, a rather paradoxical situation follows from equations (5.31) and (5.32) according to which

$$A_2 \gg A_1. \tag{5.33}$$

It means that when RPS-rippled metal surfaces are irradiated, a situation can be encountered when a significant part of the energy can be evacuated and dissipated as heat outside the irradiation spot. It appears that the energy dissipation inside the irradiation spot can become significantly weaker than dissipation outside it, if the area inside the irradiation spot is not plane but rippled ($A_{SEW} \gg A_0$) and the condition $\alpha x_0 \ll 1$ and (5.30) are fulfilled.

Some estimations are given in the following, for the case of irradiation at $\lambda \sim 10\,\mu$m of an aluminium sample the surface of which is covered, inside the irradiation spot of $x_0 = 0.5$ cm, with a diffraction grating of period $\Lambda = \lambda(\theta = 0)$. From table 5.1 we then get $A_0 = 1.1\%$ and $\alpha_d = 0.15\text{ cm}^{-1}$. By further introducing $h = h_0 = 0.16\,\mu$m and using equation (5.26), we obtain $\alpha(h = h_0) = 2\alpha_d = 0.3\text{ cm}^{-1}$. For the chosen value of x_0 we have $\alpha x_0 = 0.15 \ll 1$, and so we find ourselves in the case of small irradiation spots. And, finally, from equations (5.31) and (5.32) it follows that

$$A_1 = 0.75\% \qquad \text{and} \qquad A_2 = 15\% \tag{5.34}$$

i.e. we face a clear size effect for the experimental observation of which condition (5.30) also has to be fulfilled. In the opposite case, the value of A_1 does not change, while A_2 can be significantly smaller.

We emphasise that A_1 takes rather small values; indeed, according to equation (5.31), even for $G(h) = 1$ this quantity is no larger than $A_1 \sim 0.01$. One can also obtain an analytical expression for the A_1 maximum in the case of small irradiation spots:

$$A_1^{\max} \simeq 0.76(\alpha_d x_0) \qquad \text{when } h = h_0\left(\frac{1.9}{\alpha_d x_0}\right)^{1/2} \qquad \alpha x_0 \ll 1. \tag{5.35}$$

The considerations above are in agreement with several available experimental results.

An evident size effect was observed in reference [182] for TEA CO_2 laser irradiation of some aluminium samples in vacuum. The samples were plates of rectangular shape with dimensions largely differing in width, y_T and length, x_T, $x_T \gg y_T$. The irradiation spot was in fact square-shaped, of sides y_T, and was placed close to one of the target ends. It was noted that whenever the target melting threshold was crossed, when RPS were forming on the target surface (the microscopical investigations of the irradiation zone showed the existence of RPS 'frozen' after the laser action), the target absorptivity, A, would depend on the orientation of the radiation polarisation to the incidence surface. More precisely, it was shown that A is larger when the $\boldsymbol{E}_0$ vector of the incident wave was directed along the longer side of the rectangular target. In this case the RPS grooves are oriented in such a way that SEW are also induced and propagate along the longer side of the target, so that an important part of their energy can be transferred to the target outside the irradiation spot—a larger share as compared with the case when $\boldsymbol{E}_0$ is oriented along the shorter side of the target (see figure 5.9). On the other hand no dependence of the absorptivity A on the orientation of the vector $\boldsymbol{E}_0$ was observed for irradiation of square-shaped samples when the irradiation spot was placed in the centre of the sample.

The correctness of the examined model was also sustained by the measurements reported in reference [183] of the angular dependence of absorptivity for a sample prepared from an aluminium alloy, whose surface was covered with a diffraction grating having a rectangular profile of period

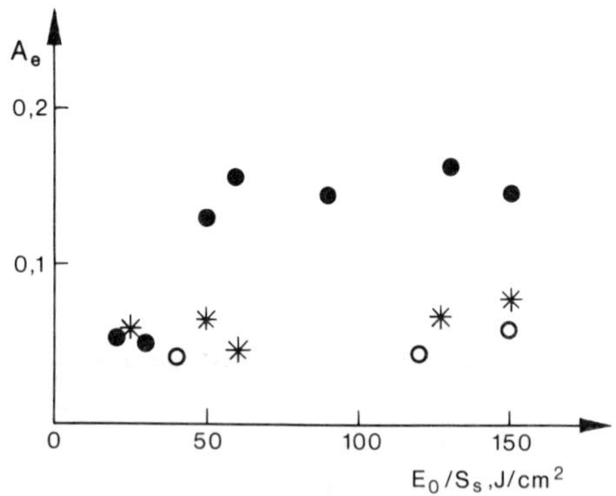

Figure 5.9 The experimental dependence of the effective absorptivity of an aluminium sample on the laser fluence in the radiation pulses generated by a TEA CO_2 laser source. The (○) points correspond to a square-shaped sample when the irradiation spot was placed in the centre of the sample. Alternatively, in the case of a sample of rectangular shape, the irradiation was performed by orienting the $\boldsymbol{E}_0$ vector of the electrical field of the light wave parallel to the longer side of the sample (●) and perpendicular to it (∗), respectively.

$\Lambda = 13.6\ \mu$m and depth $h = 1.5\ \mu$m. The radius $x_0 = 3.5$ mm was chosen close to the sample radius of $x_T \simeq 5$ mm. The grating grooves were oriented perpendicularly to the $\boldsymbol{E}_0$ vector. A micrograph of the grating surface is shown in figure 5.10 along with the experimentally obtained $A(\theta)$ curve. One can see that close to the angle $\theta_1 = \theta_r = 12.5°$, when the condition $\Lambda = \lambda/(1 - \sin\theta)$ for the SEW excitation and propagation is met, A resonantly increases from $A_0 \simeq 0.029$ to $A_1 \simeq 0.15$.†

For comparison, let us look at some numerical evaluations of the quantity $A \simeq A_0 + A_{SEW}$. Corresponding to the experimentally measured $A_0 = 0.029$ and an incidence angle $\theta_1 = \theta_r = 12.5°$, one gets from formula (5.25) $h_0 \simeq 0.365\ \mu$m. By extracting the sinusoidal harmonics of amplitude h_{ef} from the rectangular profile of the grating, we get

$$h_{ef} \simeq \left| \frac{2}{\Lambda} 2 \int_0^{\Lambda/2} h \sin\left(\frac{2\pi}{\Lambda} x\right) \mathrm{d}x \right| \simeq \frac{4h}{\pi} \simeq 1.91\ \mu\mathrm{m} \tag{5.36}$$

i.e. $h_{ef} \gg h_0$, and consequently the situation $A_1 = 1$ cannot be reached in this case.

By taking for the aluminium alloy $\alpha_d \simeq 0.5\ \mathrm{cm}^{-1}$, we obtain a SEW free path $l_{SEW} \simeq 1/2\ \alpha_d \simeq 1$ cm, close to the target dimension, i.e. one can ignore the size effect and we have $A \simeq A_0 + A_1$, to be measured experimentally. The same conclusion also follows from the evaluation of αx_0, as for the given value of h_{ef} and x_0 from equation (5.26) we obtain $\alpha x_0 \simeq 9.93$. Accordingly, from equations (5.22)–(5.24) one gets $F_1(\alpha x_0) = 0.85$, $F_2(\alpha x_0) = 0.05$ and $G(h_{ef}) = 0.136$. Then with the aid of equations (5.20) and (5.21) we obtain $A_2 \ll A_1$ and $A_0 + A_1 \simeq 0.15$—in excellent agreement with the experimentally measured value of A (the first resonance, see figure 5.10(*b*)).

Ending this section, let us consider the influence on the absorptivity of the sample of small deviations $\delta\theta$ of the incidence angle from the angle θ_r of SEW resonant excitation.

For $\delta\theta \ll \theta_r$, the SEW field can be written in the form

$$a(x, \theta_r + \delta\theta) = \exp(-\alpha x) \int_{-\infty}^{x} \exp(\alpha\xi) f(\xi, y) \tilde{\beta}(\xi)\, \mathrm{d}\xi \tag{5.37}$$

where

$$\tilde{\beta}(\xi) = \beta \exp\{ik[(\sin\theta_r + \delta\theta) - \sin\theta_r]\xi\}. \tag{5.38}$$

Assuming as before a uniform distribution of energy within the irradiation spot, we get, after several transformations and simplifications, for large

† The second maximum of the curve $A(\theta)$ corresponds to propagation along the grating surface of the radiation diffracted in the second order at an angle θ_2.

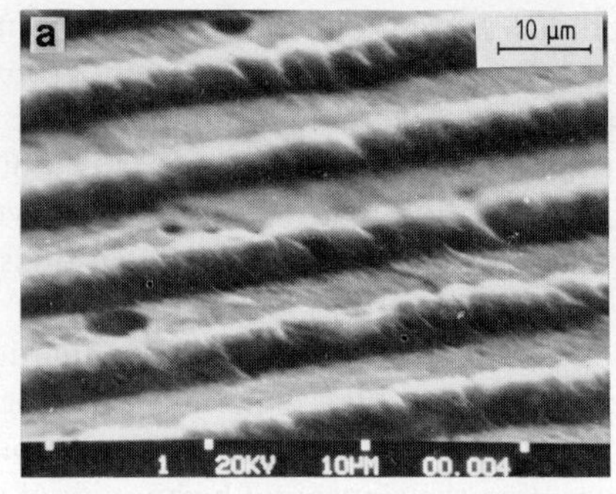

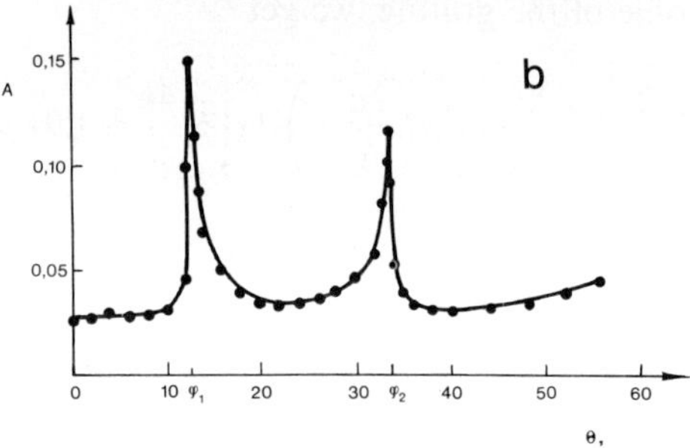

Figure 5.10 (*a*) A micrograph of the diffraction grating. (*b*) The dependence of absorptivity on the incidence angle of the radiation with $\lambda = 10.6\ \mu$m.

irradiation spots ($\alpha x_0 \gg 1$)

$$\frac{A_1(\theta_r + \delta\theta)}{A_1(\theta_r)} = \frac{1}{1 + (\delta\theta/\delta\theta_a)^2} \tag{5.39}$$

i.e. the evolution $A(\theta_r + \delta\theta)$ has, with the given hypotheses, a Lorentzian shape with an angular halfwidth

$$\delta\theta_a \simeq \frac{\alpha}{k\cos\theta} \tag{5.40}$$

which is determined by the angular dependence of the plasmon resonance and which increases quadratically (through the intermediary of α) with the increase in h (see also equation (5.26)).

For small irradiation spots $\alpha x_0 \ll 1$, and when the conditions $(kx_0 \cos\theta_r \delta\theta)^2 \gg (\alpha x_0)^2$ and $4\sin^2(kx_0\cos\theta_r\delta\theta/2) \gg (\alpha x_0)^2$ are also fulfilled, the following relation can be inferred

$$\frac{A_1(x_0, \theta_r + \delta\theta)}{A_1(x_0, \theta_r)} \simeq \frac{A_2(x_0, \theta_r + \delta\theta)}{A_2(x_0, \theta_r)} \simeq \frac{\sin^2\psi}{\psi^2} \tag{5.41}$$

where

$$\psi = kx_0 \cos\theta_r \delta\theta/2. \tag{5.42}$$

Let us now compare the halfwidth of the functions $A_1(\theta)$ and $A_2(\theta)$ with the halfwidth of the angular diffraction divergence of the radiation

$$\delta\theta_b = \frac{\lambda}{2x_0 \cos\theta}. \tag{5.43}$$

From equations (5.40) and (5.43) we get

$$\frac{\delta\theta_a}{\delta\theta_b} = \frac{\alpha x_0}{\pi} \tag{5.44}$$

wherefrom it follows that the angular divergence of the beam falling on the sample's surface, characterised by quantity $\delta\theta_b$, plays an essential role in determining the angular dependence of the absorptivity in the case of small irradiation spots, while having an insignificant influence on it for large irradiation spots.

5.5. Effects caused by RPS on metal surfaces

5.5.1. The temperature dependence of the absorptivity of rippled surfaces

Let us examine the case of a rippled surface uniformly heated by radiation, when the temperature excursion between the interference maxima and minima of SEW and the incident beam is cancelled by heat conduction into the material. It is not difficult to obtain the corresponding condition for the irradiation duration $\tau_p \gg (\Lambda/2)^2/\kappa$ (where κ is the heat diffusivity of the metal). Under these conditions, the sample absorptivity can change in quite different ways —determined by the choice of values of the parameters θ, h and x_0—when the sample temperature changes as an effect of laser irradiation.

Firstly, the width of the resonance contour $A(\theta_r + \delta\theta)$ changes by heating

$$\Delta\theta = \cos^{-1}\theta_r \left[\Delta\eta + \frac{\lambda}{\Lambda}\left(\frac{\Delta\Lambda}{\Lambda}\right)\right] \tag{5.45}$$

where $\eta = k_{p_0}/k$, and k_{p_0} is the acting part of the wavevector of SEW. For the estimation of η when $h < h_0$, one can use the expression

$$\eta = 1 + \tfrac{1}{2}\left(\frac{k^2 - n^2}{k^2 + n^2}\right). \tag{5.46}$$

One can easily show that within the 1–10 μm wavelength range, the modification of $\Delta\eta$ as a result of the temperature dependence of the optical properties of the metal is small in comparison with the second term in the right-hand side of equation (5.45), a term that is determined by the modification of the RPS period as an effect of the heat-induced expansion of the metal

$$\Delta\Lambda = \Lambda\alpha_T(T - T_0) \tag{5.47}$$

where α_T is the coefficient of linear expansion of the metal.

In this way one obtains from (3.45) for $\theta_r = 0$ ($\Lambda = \lambda$), a rather simple relation

$$\Delta\theta \simeq \alpha_T(T - T_0). \tag{5.48}$$

If, along with the increase in the quantity $T - T_0$, $\Delta\theta$ gets closer to or even exceeds the resonant halfwidths $\delta\theta_a$ or $\delta\theta_b$ (as found from equations (5.40) and (5.43)), the rippled surface, generally speaking, no longer behaves resonantly for SEW excitation. This would lead to a marked decrease in the quantity A_{SEW} and of the results obtained when measuring the total absorptivity, determined from

$$A(T) = A_0(T) + A_{SEW}(T). \tag{5.49}$$

The calculation shows that for this effect to appear a temperature $T \geqslant 100\,^\circ C$ is required. We also mention that the amplification of the absorptivity A_{SEW} by heating as a result of the displacement of the contour $A(\theta_r)$, can appear if the grating is initially not in resonance but then due to the light-induced heat expansion and the increase in $\Delta\theta$ it becomes resonant to SEW excitation and propagation.

Secondly, apart from the changes in $\Delta\theta$, the change of $A_{SEW}(\theta = \theta_r)$ by heating proceeds from the dependence $h_0(T)$.

To begin with, let us examine the case of large irradiation spots. Then, assuming that $h(T) \simeq$ const, from equations (5.20), (5.24) and (5.25) we obtain

$$A_{SEW}(T) = A_1(T) = \frac{4h^2h_0^2}{(h_0^2 + h^2)^2} \tag{5.50}$$

$$h_0^2(T) = A_0(T)\cos\theta/g \tag{5.51}$$

where $g = 2\pi/\Lambda$. From equations (5.50) and (5.51) it follows that, in contrast with $A_0(T)$ which increases monotonically with the temperature, $A_{SEW}(T)$ can be, depending on the h depth value, either an increasing or a decreasing function of temperature. We can identify three ranges of h values, featuring different evolutions of the curves $A_1(T)$.

When $h^2 \ll h_0^2(T_0) \ll h_0^2(T_f)$—meaning that the gratings are of small amplitude—the quantity $A_1(T)$ decreases monotonically with temperature over the whole range from room temperature to the final temperature, T_f (see curve 1 in figure 5.11)

$$A_1(T) = \frac{4h^2g^2}{A_0(T)\cos\theta}. \tag{5.52}$$

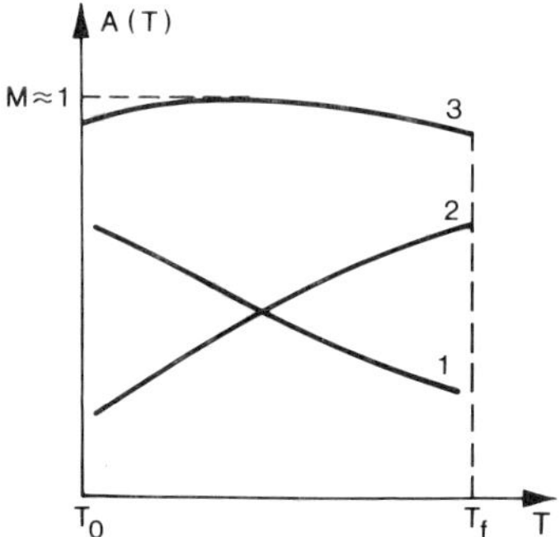

Figure 5.11 The characteristic curves of the temperature dependence of the absorptivity $A_1(T)$ of a rippled surface for different h values: (1) $h^2 \ll h_0^2(T_0) \ll h_0^2(T_f)$; (2) $h^2 \gg h_0^2(T_0), h_0^2(T_f)$; (3) $h_0^2(T_0) \leqslant h^2 \leqslant h_0^2(T_f)$.

Conversely, for a deep grating ($h^2 \gg h_0^2(T_0)$, $h_0^2(T_f)$), the quantity $A_1(T)$ increases with temperature (curve 2 in figure 5.11) and the equation for $A_1(T)$ has the form

$$A_1(T) = \frac{A_0(T)\cos\theta}{4h^2 g^2}. \tag{5.53}$$

And, finally, in the intermediate case $h_0^2(T_0) \leqslant h^2 \leqslant h_0^2(T_f)$ the quantity $A_1(T)$ is close to the maximum $A_1 \simeq 1$, and changes with temperature as one can see from figure 5.11 (curve 3).

These features of the curves $A_1(\theta, T)$ can also influence in a significant way the temperature dependence of the total absorptivity, $A = A_0 + A_1$, of a metal sample with a rippled surface (rough, though exhibiting clearly expressed resonant spatial harmonics).

The analysis of the $A(T)$ dependence in the case of small irradiation spots ($\alpha x_0 \ll 1$) becomes even more difficult when one has to take into account the size effects by heat dissipation into the sample outside the irradiation spot.

5.5.2. *The field amplification in gas (vacuum) close to a metal surface with RPS*

We shall examine now the ratio of the intensity of SEW (we shall denote this quantity by I_{SEW}) to the intensity I_0 of the light wave incident on RPS.

Within the limits of the irradiation spot, $x \leqslant x_0$, we have

$$\frac{I_{SEW}}{I_0} = \left|\frac{a}{f}\right|^2 = \frac{4\cos\theta}{A_0}\frac{4h^2h_0^2}{(h_0^2+h^2)^2}[1-\exp(-\alpha x)]^2. \tag{5.54}$$

From equation (5.54) two main conclusions immediately follow. First the maximum value of the ratio I_{SEW}/I_0 is reached at the border $x = x_0$ of the irradiation spot. Secondly, the maximum amplification in the ratio I_{SEW}/I_0

is to be obtained for $h = h_0$ and $\alpha x_0 \gg 1$

$$\left(\frac{I_{SEW}}{I_0}\right)_{max} = \frac{4\cos\theta}{A_0}. \tag{5.55}$$

From equation (5.55) one can see that the greatest amplification effect can be obtained for perpendicular incidence of the radiation, while the $(I_{SEW}/I_0)_{max}$ values can reach ~ 40–400 for a typical value of the initial absorptivity of metals, $A_0 \sim 0.01$–0.1.

Accordingly, in the case of large irradiation spots one can expect amplification of the radiation intensity on the rippled metal surface by one to two orders of magnitude. The region of field amplification is determined by the attenuation depth of SEW, δ_{SEW}, along the perpendicular on the irradiated surface, i.e.

$$I_{SEW}(z) = I_{SEW}(z=0)\exp(-z/\delta_{SEW}) \tag{5.56}$$

where z is the distance from the sample surface and δ_{SEW} is to be obtained from equation (5.15). As an example, for aluminium, with $\lambda = 1\ \mu$m and $\lambda = 10\ \mu$m, δ_{SEW} is equal to 1 μm and 100 μm, respectively.

Obviously, such a marked amplification of the field over a rather extensive region close to the sample can significantly influence the processes that evolve close to the irradiated, rippled (rough), surface. Such an effect should appear, among others, in the process of plasma formation on a solid sample, for although the target surface is initially plane, it is soon covered with RPS during melting and vaporisation under the action of laser radiation.

In the case of small irradiation spots ($\alpha x_0 \ll 1$) the effect of field amplification is less significant for the same depth of grating. The ratio I_{SEW}/I_0 is reduced by a factor of $(\alpha x_0)^2$.

Let us illustrate the difference between small and large irradiation spots in the case of aluminium for $\lambda = 10\ \mu$m, $\theta = 0$, $h = h_0 = 0.16\ \mu$m and $T = T_0$. Then for $x_0 = 0.2$ cm ($\alpha x_0 \ll 1$) we have $(I_{SEW}/I_0)_{max} \simeq 1.3$ while for $x_0 = 20$ cm ($\alpha x_0 \gg 1$) the maximum amplification of intensity is $(I_{SEW}/I_0)_{max} \simeq 360$.

In ending this section, we mention that the amplification of the electromagnetic field on the periodic relief was also studied in connection with an investigation of the prospects for using the diffraction gratings for obtaining maximal signals of coherent scattering of light by the molecules adsorbed on the rough surface of the metal (see, for example, references [184, 185]). These calculations were, however, performed by numerical methods, and an optimisation of the problem was not possible.

5.5.3. The spatial modulation of absorptivity by short laser pulses

It is a fact that even when the relief depth is small ($h \ll h_0$) and, correspondingly, the supplementary absorptivity A_1 is also small (even much smaller than A_0) several interesting effects occur which are determined by the RPS.

As an effect of the spatial modulation of radiation on the rippled surface, local values of absorptivity which differ from each other appear in the interference maxima, A_l^+, and minima, A_l^-. For $x \leqslant x_0$ their values result from

$$A_l^{\pm}(x, h) = \left[1 \pm \left(\frac{\cos\theta}{A_0} \right)^{1/2} \frac{2hh_0}{h^2 + h_0^2} [1 - \exp(-\alpha x)] \right]^2 \frac{A_0}{\cos\theta} \quad (5.57)$$

while their difference is

$$A_l^+(x, h) - A_l^-(x, h) = \frac{8hh_0}{h^2 + h_0^2} [1 - \exp(-\alpha x)] \left(\frac{A_0}{\cos\theta} \right)^{1/2}. \quad (5.58)$$

One can also obtain an analytical expression for the local absorptivity averaged over the period Λ of the structure

$$\begin{aligned} \bar{A}_l(x, h) &= \tfrac{1}{2}[A_l^+(x, h) + A_l^-(x, h) \\ &= \left(1 + \frac{\cos\theta}{A_0} \frac{4h^2h_0^2}{(h^2 + h_0^2)^2} [1 - \exp(-\alpha x)]^2 \right) \frac{A_0}{\cos\theta} \end{aligned} \quad (5.59)$$

which can be related to $A_1(x, h)$ by the following expression ($x = x_0$)

$$\frac{1}{x_0} \int_0^{x_0} \bar{A}_l(x, h)\, \mathrm{d}x = A_1(x_0, h) + \frac{A_0}{\cos\theta}. \quad (5.60)$$

The characteristic curves $A_l^{\pm}(x, h)$ and $\bar{A}_l(x, h)$ are presented in figure 5.12 as functions of x, for three values of the relief depth, h.

Equations (5.57)–(5.59) are much simpler in the case of large irradiation spots ($\alpha x_0 \gg 1$) and a small relief ($h^2 \ll h_0^2$). Then the dependence of local absorptivity on $x \leqslant x_0$ disappears and the corresponding expressions take the form

$$A_l^{\pm}(h) = \left(1 \pm \frac{h}{h^*} \right)^2 \frac{A_0}{\cos\theta} \quad (5.61)$$

$$A_l^+(h) - A_1^-(h) = \frac{4h}{h^*} \frac{A_0}{\cos\theta} \quad (5.62)$$

$$\bar{A}_l(h) = [1 + (h/h^*)^2] \frac{A_0}{\cos\theta} = A_1 + \frac{A_0}{\cos\theta} \quad (5.63)$$

where the characteristic depth h^* of the relief is to be determined from equation (5.29). We mention that along with the increase in h (within the limits $h^2 \ll h_0^2$), A_l^+ increases monotonically and A_l^- begins by decreasing down to a minimum $A_l^- = 0$ before increasing again. However, even more important is the fact that along with the amplification of h we observe increasingly significant differences between the values of the local absorptivity in the maxima and minima of the interference pattern, while the absorptivity

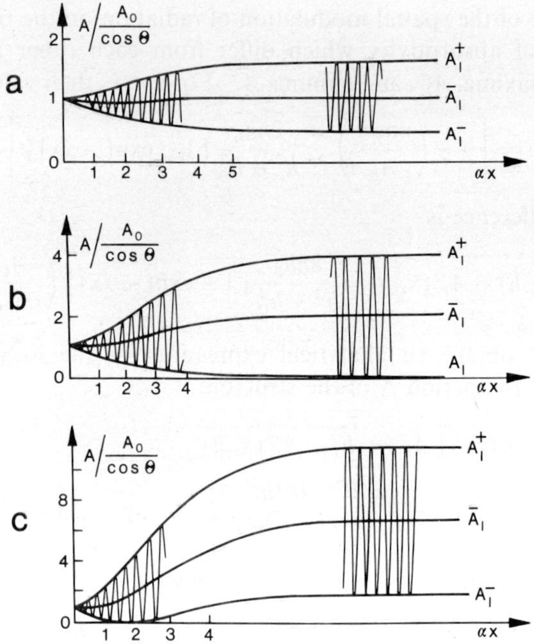

Figure 5.12 The variation along the surface of the diffraction grating of the absorptivities—average $\bar{A}_l$ and local in the interference maxima A_l^+ and minima A_l^-, of the incident radiation and SEW, respectively—for (*a*) $h < h^*$, (*b*) $h \simeq h^*$ and (*c*) $h > h^*$.

averaged over the spot remains practically unchanged, and it is precisely the value of $\bar{A}_l$ which is experimentally determined.

We shall assume $h = 0.1h^*$—which, according to table 5.1, corresponds for $\lambda = 10\,\mu\text{m}$ to a relief depth of only ~ 10 Å. Then, from equations (5.61)–(5.63) we obtain $A_l^\pm = 1.21 A_0(\theta)$, $A_l^- = 0.81\ A_0(\theta)$ and $A_l^+/A_l^- \simeq 1.5$, i.e. the differences in the local absorptivity would not suffice to influence in any significant manner the various processes that evolve on the surface of the irradiated sample. In this case $A_1 = 0.01\ A_0(\theta)$ and, correspondingly, $\bar{A}_l = 1.01\ A_0(\theta)$. Such modifications of the absorptivity averaged over the irradiation spot can of course be considered as negligibly small.

The depth of the spatial modulation of heat dissipation into the sample, starting with a grating amplitude of $0.1h^*$, can prove important, and in some cases is a decisive factor in the processes of laser heating of surfaces (laser damage, laser stimulation of surface chemical reactions, etc). It is therefore necessary that sufficiently short radiation pulses with $\tau \ll (\Lambda/2)^2/\kappa$ are used so that the temperature excursion $T^+ - T^-$ between the maxima and the minima of the interference pattern is not extinguished by the heat conduction mechanism. The temperature gradients appearing in this case along the surface

$$\frac{dT}{dx} \simeq \frac{T^+ - T^-}{\Lambda/2} \tag{5.64}$$

can reach huge values of $\sim 10^5$–10^6 K cm^{-1} for radiation with $\lambda = 10.6$ μm, and $\sim 10^6$–10^7 K cm^{-1} for radiation with $\lambda = 1$ μm. In this case the average temperature of the surface

$$< T > = \tfrac{1}{2}(T^+ + T^-) \tag{5.65}$$

stays rather low: $< T > = 50$–500 °C, which is usually lower than the melting point.

For such large temperature gradients, which can become even larger with the shortening of λ, the heat and charge transfer processes on the surface can change significantly—which opens up new prospects for the use of lasers in investigations of surface phenomena [186]. We also mention that the existence of large dT/dx can influence the very process of SEW generation.

Analogously to the previously examined case of large irradiation spots an analytical expression can also be obtained for small irradiation spots, $\alpha x_0 \ll 1$

$$A_l^{\pm}(x, h) = \left(1 \pm \alpha_d x \frac{h}{h^*}\right)^2 \frac{A_0}{\cos\theta} \tag{5.66}$$

according to which $A_l^{\pm}$ becomes dependent on x (according to equations (5.66) and (5.67)). An example of calculation in the case of radiation with $\lambda = \Lambda = 10$ μm ($\theta = 0$), for an aluminium sample with $T = T_0$, $x_0 = 1$ mm and two h values is given in table 5.2.

The models applied to the analytical calculation of the local absorptivity of a rippled surface can also be used in investigations of the dynamics of RPS development. For example, for the mechanism of relief deepening entirely by vaporisation, and neglecting melt removal under the action of a reactive recoil pulse of vapour, one can obtain the following expression for the change in the total amplitude $2h$ of RPS in time

$$2\frac{dh}{dt} = \frac{A_l^+(x, h) - A_l^-(x, h)}{\rho q} I_0 \cos\theta \tag{5.67}$$

Table 5.2 The local, A_l^+, A_l^- and average $\bar{A}_l$ values of absorptivity in case of $\lambda = 10$ μm laser irradiation of an aluminium sample.

h (μm)	A_l/A_0	$\bar{A}_l/A_0$	A_l^+/A_0	A_l^-/A_0	A_l^+/A_l^-
0.15	0.02	1.06	1.54	0.56	2.6
0.6	0.33	2	4	0	∞

where ρ and q are the density and the specific heat of vaporisation, respectively of the sample material.

The solution of equation (5.67) for the case of large irradiation spots $\alpha x_0 \gg 1$, $h \ll h_0$ and $\theta = 0$ takes the form

$$h(t) \sim \exp(\gamma t) \tag{5.68}$$

where

$$\gamma = \frac{2A_0 I_0}{\rho q h^*} = \frac{8\pi I_0}{\rho q \Lambda}. \tag{5.69}$$

The fact that the quantity γ is positive points to the existence of a positive feedback in the problem of RPS development, i.e. irradiation at small h is accompanied by relief deepening.

By assuming $\gamma t = \gamma \tau_p = 1$, one can infer the expression for the threshold laser pulse fluence for the growth of RPS

$$E_s^p = I_0^p \tau_p = \frac{\rho q \Lambda}{8\pi}. \tag{5.70}$$

Once more for the case of irradiation of RPS on an aluminium sample ($\rho = 2.35\ \mathrm{g\,cm^{-3}}$, $q = 10^4$ J) we obtain from equation (5.70), $E_s^p \simeq 1\ \mathrm{J\,cm^{-2}}$ for $\lambda = 10\ \mu$m, while for $\lambda = 1\ \mu$m we obtain $E_s^p \sim 0.1\ \mathrm{J\,cm^{-2}}$. For this, as said before, the condition $\tau_p \ll \lambda^2/4\kappa$ has to be fulfilled.

Finally, we emphasise again that all the methods and approaches we have dealt with for the calculation of the RPS parameters of an ideally wavy surface and also the consideration of their influence upon the interaction processes of laser radiation with samples can also be applied for simple, rough surfaces, whenever these feature spatially resonant roughness harmonics.

Chapter 6 Laser-induced Oxidation and Burning of Metals

This chapter discusses processes of thermochemical origin developing when metals are exposed to laser irradiation in chemically active environments (e.g. oxygen, air etc). In such circumstances reaction products can accumulate in quantities large enough to influence, or even completely determine, the very course of the process. Two such effects are considered: the modification of the optical properties of the irradiated target, and the additional heat release due to exothermic reactions.

6.1. Mechanisms resulting in the increase of the thermal effect of CO_2 laser radiation upon metals in an oxidising medium

The reaction products which grow during laser irradiation of solid surfaces in contact with a chemically active gas may reach a thickness and have a growth rate sufficiently high to result in a significant change in the radiation–target interaction. Many situations occur when the superficial chemical changes are not determined by the radiation monochromaticity, meaning that their character is non-resonant, changes then being induced by target heating under the action of laser radiation.

A typical example of this kind of interaction, which has been called 'thermochemical', is the heating of an oxidising metal. Beginning with the first work in this field [187–196] two processes have been mentioned which accompany metal heating in an oxidising atmosphere (air, oxygen) and may result in a considerable increase in the effect of CO_2 laser radiation on metal targets which, under normal conditions, strongly reflect 10.6 μm radiation.

First, the absorption coefficients of the majority of metal oxides within this wavelength range are high, showing values of $\sim 10^2$–$10^4\,cm^{-1}$. If the oxide layer that forms on the metal surface during the initial irradiation stages is sufficiently thick, the absorptivity, A, of the metal–oxide system may considerably exceed the metal absorptivity, A_M (even if its temperature variation $A_M(T)$ is taken into consideration), and subsequent sample heating is performed under conditions of a higher absorptivity, $A \gg A_M$. That is why the process of prolonged laser heating of an oxidising metal sample (for example under the action of CW laser radiation) proceeds in two steps. In the first step, when the oxide film is thin the heating is relatively slow. The second step is connected to the activation of the oxidation reaction.

Considering that oxidation, like any other chemical reaction, depends exponentially upon temperature, a considerably higher growth rate of oxide film thickness is to be expected beyond a certain temperature, T_a.† Accordingly, a rapid increase in absorptivity occurs which, in turn, leads to an increase in the rate of energy storage in the target and to the amplification of the rate of variation of temperature in the interaction spot. The occurrence of such a positive feedback between heating, oxidation and absorptivity increase may result, with $T > T_a$, in a cascading increase in the efficiency of thermal processing of the metal under the action of the laser beam.

Another characteristic of the thermochemical interaction mechanism consists in a supplementary energy release into the target during the evolution of the oxidation reaction (owing to its exothermal nature). The higher the metal temperature, the higher is the oxidation reaction rate and, implicitly, the generated power, P_{ex}. If P_{ex} becomes comparable to or exceeds AP, laser ignition or burning of the metal occurs. It is evident that such conditions of thermal action on the metallic target can be obtained only with high enough values of the temperature, T, i.e. only in the activation stage of the oxidation.

The idea of using thermochemical surface reactions to increase the efficiency of thermal processing of metals under the action of CW CO_2 lasers belongs to Sullivan and Houldcroft [187]. Circulating a chemically active gas (oxygen) in the laser irradiation area was proposed in order to increase the absorptivity of the target by oxidation. Moreover, an analogy was made with the traditional methods of metal cutting, in which the oxygen jet gives way to the metal combustion regime at high temperatures, also removing the reaction products from the interaction area. This processing method was called the gas jet laser cutting (GJLC) technique [187].

It has to be stressed that the multifunctional action of the oxygen jet (convective heat removal from the clearing area is also to be considered as well as the above-mentioned effects) makes it considerably more difficult to understand the role of the thermochemical mechanism in reducing the laser energy consumption under a GJLC regime. As a consequence, the main

† For the majority of metals T_a is less than the melting point T_m.

goal of the research conducted in this field has for a long time been (e.g. [2, 3, 105]) to empirically establish the processing regimes (not always optimal) by varying the intensity of CW CO_2 laser radiation as well as the parameters of the gas jet and the speed of the moving parts. The fundamental laws governing the thermochemical effect of laser radiation with GJLC has remained, however, an untackled topic.

It is evident that any analysis regarding the process of laser heating of an oxidising metal is possible only on the basis of a sound knowledge of the laws of the respective chemical kinetics. The specificity of laser oxidation consists in the fact that the chemical reaction develops under the action of a heat source of variable power (owing to the variation of sample absorptivity with the thickness and contents of the oxide layer and with sample temperature) under non-isothermal conditions. Also, the most important part of the traditional classical research regarding the kinetics of metal oxidation was conducted under rigorous isothermal conditions (see, for example, [197–201]). The thorough studies on high-temperature heating and oxidation of a series of metals, which have been initiated over the past years, have shown that the chemical kinetics is considerably complicated under conditions of a variable sample temperature and depends on the heating procedure [202].

For this reason it was necessary to undertake both experimental and theoretical research on the kinetics of laser oxidation and heating of metals under model assumptions which are as streamlined as possible.

This chapter is dedicated to the analysis of the main results of this research conducted during the last 5–10 years (selective areas of this research field have also been reviewed in references [202–205]). There has been a dramatic increase in the number of publications dedicated to these topics and significant progress has been made in the understanding of the thermochemical mechanism of the interaction of radiation with metals.

6.2. Fundamental equations

The physical fundamentals of the thermochemical mechanism of oxidation of metals by heating were first examined in references [191–194], being subsequently amended and clarified. At the bottom of it all, is the concurrent solution of three basic equations: (i) the kinetic equation for the variation of oxide film thickness x with temperature T reached on the irradiated surface of the target; (ii) the equation describing the dependence of absorptivity on x; and (iii) the equation of thermal conduction, connecting T and A.

(i) It is well known (see, for example, references [197–202]) that the formation of the oxide layer evolves in several phases, comprising oxygen admission and absorption into the surface, metal and oxygen transfer across the oxide layer towards the interphase boundary, and the chemical reaction leading to the formation of new compounds. The very complex character of the oxidation process expresses the fact that the laws of chemical kinetics do not depend on the nature of the metal only but also on the reaction conditions. In particular, the oxidation rate $\mathrm{d}x/\mathrm{d}t$ is determined by the slowest (most limiting) phase of the process.

To start with, let us examine the simplest and most common case, i.e. when the reaction rate is limited by the diffusion of the reaction components through the oxide layer. If the oxide forms a compact film on the metal surface, a spatial separation of the participants in the reaction $n\mathrm{Me} + \frac{1}{2}m\mathrm{O}_2 \rightarrow \mathrm{Me}_n\mathrm{O}_m$ occurs, during which a diffusion of metal and oxygen ions through vacancies, grain boundaries and other faults of the oxide layer is mainly observed. In this case, it is essential to note that the first to diffuse are not the neutral atoms, but the oxygen and metal ions. The effect is that the reaction mechanism changes as a function of oxide layer thickness. For thin oxide layers a strong influence on the diffusion process is exerted by the electric fields appearing in the boundary layer. The theory of growth of a thin oxide film layer was given by Cabrerra and Mott [197–199, 201]. In this case the oxidation kinetics is described by the equation

$$\frac{\mathrm{d}x}{\mathrm{d}t} = v_0 \exp\left(-\frac{T_\mathrm{d}}{T}\right)\left[\exp\left(\frac{x_\mathrm{k}(T)}{x}\right) - \exp\left(-\frac{x_\mathrm{k}(T)}{x}\right)\right]. \tag{6.1}$$

Here v_0 is close to the sound velocity, T_d is the diffusion activation temperature, T is the temperature of the metal surface, x_k is the characteristic thickness of the oxide layer against which the influence of the electric field is appreciated as either strong (with $x < x_\mathrm{k}$) or weak (with $x > x_\mathrm{k}$). At room temperature $x_\mathrm{k} \sim 100$ Å, while its magnitude decreases with T [201].

However, corresponding to the absorption coefficients of the 10.6 μm radiation by metallic oxides, $\alpha \sim 10^2$–10^4 cm^{-1}, the additional radiation absorption within very thin film layers is very small ($A \approx A_\mathrm{M}$) and the formation of these layers does not influence the dynamics of metal heating by irradiation.

For an oxide layer thickness $x > 1000$ Å, when important modifications of A as against A_M and corresponding increases in the metal sample heating ratio are noted in most practical cases, the Wagner theory [197] can usually be applied, and oxidation follows a parabolic law

$$\frac{\mathrm{d}x}{\mathrm{d}t} = \frac{d}{x} \qquad d = d_0 \exp\left(-\frac{T_\mathrm{d}}{T}\right). \tag{6.2}$$

For metal oxides the characteristic values of T_d are $\sim (1$–$3) \times 10^4$ K, while the parameter d_0 depends upon the partial oxygen pressure at the metal–oxide boundary, p_{O_2}.

At sufficiently high values of chemical reaction temperature and rate, the oxidation process may not be limited by the diffusion of the reaction participants over the oxide layer, but by the admission of oxygen into the reaction zone. Such a situation is typical for the regimes of metal ignition and burning. Consequently, on the assumption that $p_0 \sim p_{O_2}$, in the simplest case of the stationary solution to the spherically symmetric problem, Wagner's parabolic law takes the form [206, 207]

$$\frac{dx}{dt} = \frac{(d_0/x)\exp(-T_d/T)}{1+\mu(d_0/x)\exp(-T_d/T)} \qquad \mu = \frac{r_{eff}}{D_{O_2}} \frac{\rho'_{O_2}}{\rho_{O_2}} \tag{6.3}$$

where r_{eff} is the effective radius of the oxidising surface, D_{O_2} the diffusion coefficient for the oxygen molecules and ρ_{O_2} and ρ'_{O_2} the values of the oxygen densities corresponding to the gas and oxide, respectively. From (6.3) it results that the law (6.1) is valid, provided that the temperature values are not very high, and $\mu(d_0/x)\exp(-T_d/T) \ll 1$, while in the opposite limit case, the equation $dx/dt \approx (D_{O_2}/r)(\rho_{O_2}/\rho'_{O_2})$ holds true. The value of the factor which influences the metal heating ratio can be reduced by blowing the irradiated area with oxygen or air jets (similarly to GJLC of metals).

(ii) In order to obtain a correct dependence $A(x)$ it is absolutely necessary to take into consideration [208] the interference phenomena within the metal–oxide layer. Let us limit ourselves here to the simplest case, i.e. when a layer of a single type of oxide grows with a uniform thickness on the surface of the metal sample. In this case, the absorptivity of the sample covered by an oxide layer of thickness x is determined by the relation [209]

$$A = 1 - |R|^2 \qquad R = \frac{r_{12}\exp(-2i\varphi) + r_{23}}{\exp(-2i\varphi) + r_{12}r_{23}} \tag{6.4}$$

where

$$\varphi = \frac{\omega}{c_0} x\sqrt{\varepsilon_c} = \tfrac{1}{2}(\beta + i\alpha)x \qquad r_{12} = \frac{1-\sqrt{\varepsilon_c}}{1+\sqrt{\varepsilon_c}}$$

$$r_{23} = \frac{r_{12} - r_{13}}{r_{12}r_{13} - 1} \qquad 1 - |r_{13}|^2 = A_M.$$

Here, $\omega = 2\pi\nu$, where ν is the frequency of laser radiation, c_0 the velocity of light, r_{12} and r_{13} the amplitude coefficients of radiation reflection on oxide and metal, and the dielectric permittivity of the oxide, $\sqrt{\varepsilon_c} = n + ik$. For media characterised by a good conductivity $A_M \ll 1$ and $r_{13} = -1 + 0.5A_M(i-1)$.

In many cases of interest, the frequency of CO_2 laser radiation meets the conditions $1 - |r_{12}|^2 \gg A_M$, $x \ll 1$, $n > 1$ and, correspondingly, $\alpha \ll \beta$. In this instance, the optical thickness of the oxide film is small, $\alpha x \ll 1$, up to the first interfering oscillation of the function $A(x)$. On these assumptions, the formula (6.4) can be substantially simplified to obtain the following approximate expression [208]

$$A(x)=\frac{n^2A_M+2k(\beta x-\sin\beta x)}{n^2+(1-n^2)\sin^2(\beta x/2)}. \tag{6.5}$$

Let us mention that in a number of papers (for example [191–194, 209]) a simpler formula has been applied for the calculation of $A(x)$, namely

$$A(x)=A_M+2\alpha x \tag{6.6}$$

though without proper indications of its range of applicability. Evidently, the formula (6.6) results from the more general expression (6.5), but only under the conditions $x \gg \lambda/(4\pi n)$, $n-1 \ll 1$ and $2\alpha x \ll 1$, which are generally not fulfilled simultaneously for $\lambda = 10.6\ \mu$m and for the typical values of α and $n \simeq 1$–2.5 for metal oxides.

Another relation can be obtained from (6.5), suitable for thin oxide films, by neglecting terms of the order $(\beta x)^3$ [208]

$$A(x)=A_M+bx^2 \qquad b=\frac{4\pi^2(n^2-1)A_M}{\lambda^2}. \tag{6.7}$$

Let us illustrate the importance of a correct approximation for $A(x)$, taking as an example a thin oxide layer with a thickness $x = 5\times10^{-5}$ cm (such a thickness is typical for the stage of oxidation activation by laser heating). With $A_M = 0.02$, $n = 1.1$, $\alpha = 2.5\times10^2\ \text{cm}^{-1}$, the quantity $2\alpha x$ included in (6.6) has the magnitude 0.025, whereas the calculation performed according to equation (6.7) gives $bx^2 = 0.0025$, i.e. a value an order of magnitude lower.

(iii) In order to make the system complete the thermal conductivity equation has to be written. The heat equation has its simplest solutions in two limiting cases of interest.

When heating a semi-infinite target with an energy source of circular cross section in a stationary regime, the dependence $T(A)$ takes the form [82]

$$T=\frac{AP}{\pi k_T R_s}+T_0 \tag{6.8}$$

where P is the average radiation power, R_s is the radius of the focal spot, k_T is the thermal conductivity of the metal and T_0 the initial temperature of the target. Although apparently simple, the application of equation (6.8) in the analysis of thermochemical metal heating is extremely tedious. Indeed, as a result of the interdependence between the processes of heating, oxidation and change in the metal's absorptivity, the heating of a semi-infinite target is accompanied by a change in the spatial profile of temperature [208, 210], even under the circumstances of a uniform distribution of energy in the cross section of the laser beam, and much more in the case of Gaussian beams (more often used for practical purposes). The analysis becomes even more complicated in a non-stationary case.

Much better chances for performing a detailed theoretical analysis in comparing calculated and experimental data are offered by the modelling of a thermally thin and thermally insulated metal foil. Under these circumstances the thermal conductivity equation is replaced by an energy conservation equation of the form

$$cm\,\frac{\mathrm{d}T}{\mathrm{d}t} = AP - Q \qquad (6.9)$$

where c and m are the specific heat and the mass, respectively of the irradiated sample, and Q is the power of the thermal losses.

If we now take into consideration the exothermal character of the reaction, the A value from equation (6.9) has to be substituted by the so-called effective target absorbtivity, $A_e \rightarrow A + P_{ex}/P$, and the power released in the reaction can be obtained from the equation

$$P_{ex} = \rho S W_H\,\frac{\mathrm{d}x}{\mathrm{d}t}. \qquad (6.10)$$

Here ρ is the density of the oxide, W_H the latent heat of the reaction and S the total area where the chemical reaction takes place.

This approach to the problem of laser heating of an oxidising metal is general enough, although by taking into account the peculiarities of the process for different metals the system of basic equations (6.3)–(6.5) can be better tailored and clarified.

In order to obtain an agreement between theoretical and experimental results it is necessary to take into account the temperature (and implicitly time) dependence of the optical constants of both metal and oxide, the structure and content of the oxide film (which can be non-uniform and made up of several components), the possibility that the oxide sublimates at sufficiently high temperatures as well as a variety of other processes.

All these topics will be examined in detail in the following sections of this chapter.

6.3. Experimental methods of investigation

Significant progress in understanding the kinetics of laser heating and oxidation of metals was made possible through the application of the method of modified calorimetric measurement [208, 211, 212]. The heating of thermally thin and thermally insulated metal foils was investigated. The general experimental scheme is presented in figure 6.1.

Continuous wave CO_2 laser radiation (1) was horizontally directed through a NaCl window (2) in a reaction chamber filled with air or oxygen at different pressures ($p_0 \leqslant 1$ atm), and was focused through a NaCl lens (3) on the metal foil (4). A foil thickness of $h \sim 0.1-1$ mm was chosen, and a diameter varying in the 3–5 mm range. The irradiation spot covered almost the entire target area. The foils were insulated from the thermoconductive elements of the installation, being supported by thin wires ($\varnothing \simeq 0.1$ mm), connecting to the recording equipment the weld of the thermocouples used for temperature control. The thermocouples (chromel–alumel, chromel–copper or tungsten–rhenium) were built into the sample, or welded, on the unirradiated side.

Target heating and cooling (after interrupting the irradiation) were recorded. The signal from the thermocouple was divided into two parts: (i) one part was directed straight into one of the multibeam oscilloscope channels; (ii) another part of the signal was first passed through a differentiating RC chain (8), then through an amplifier (9) and was subsequently introduced into the second oscilloscope channel. This enabled simultaneous recording of both time variation of target temperature $T(t)$ and of its first derivative, $\mathrm{d}T(t)/\mathrm{d}t$.

The time resolution of the recording scheme was $\tau_R \simeq 0.02$ s. The time necessary to equalise the temperature of the irradiated sample $\tau_e \simeq h^2/\kappa_M$ did not exceed 0.01–0.1 s (corresponding to the metal's thermal diffusivity of $\kappa_M \simeq 0.1-1\ \mathrm{cm^2\,s^{-1}}$), while typical heating durations τ_p were chosen in the range between a few seconds and a few minutes. It follows that $\tau_p \gg \max(\tau_R, \tau_e)$, and the targets can be considered as thermally thin.

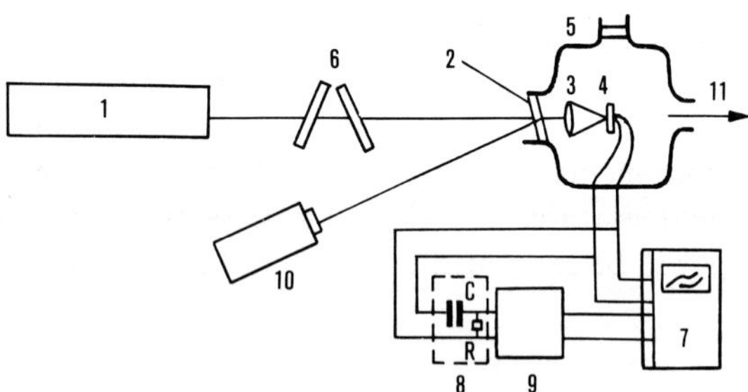

Figure 6.1 Typical experimental set-up for determining the dependence on temperature of the effective absorptivity of thermally thin metallic foils: (1) cw CO_2 laser source; (2) calibrated beam-splitter; (3) focusing lens; (4) metallic foil; (5) irradiation chamber; (6) calibrated attenuators; (7) oscilloscope; (8) *RC* differentiator; (9) amplifier; (10) power meter; (11) to the vacuum pump.

If the heat exchange with the environment is taken into consideration and $P_{ex} = 0$, the stage of foil heating (+ index) can be described by the relation

$$cm \left.\frac{dT}{dt}\right|_{+} = AP - Q_{+}(T). \tag{6.11}$$

Consequently, when the laser radiation action ceases (− index) the thermal balance equation takes the form

$$cm \left.\frac{dT}{dt}\right|_{+} = -Q_{-}(T). \tag{6.12}$$

It is easy to show that, under the conditions of simultaneous recording of the signals $T(t)$ and $dT(t)/dt$ during the process of sample heating and cooling, the evolution of the temperature- (or time-) dependent absorptivity can be established for $P_{ex} = 0$. Indeed, let us assume that within a certain time interval $\Delta t_i > \max(\tau_R, \tau_e)$, the sample temperature changes by ΔT_i and becomes equal to T_i, where $\Delta T_i \ll T_i$. In this case, subtraction of (6.12) from (6.11), for the same value of temperature, results in the following relation

$$A(T_i) = \frac{cm}{P}\left(\left.\frac{dT}{dt}(T_i)\right|_{+} - \left.\frac{dT}{dt}(T_i)\right|_{-} \right). \tag{6.13}$$

In establishing equation (6.13) it was assumed that heat losses during sample heating, $Q_{+}(T_i)$, and cooling, $Q_{-}(T_i)$, are identical, which is not true under the conditions of surface oxidation. Taking this difference into account, the actual value of absorptivity, A_R, differs from that just established by the amount

$$A_R - A = \frac{Q_{+}(T_i) - Q_{-}(T_i)}{P}. \tag{6.14}$$

Under conditions of laser heating of a thermally insulated foil, the total power losses, Q, by re-radiation, convection and thermal conduction through thermocouples and connection wires (Q_T), are given by the expression

$$Q = S\sigma_{SB}\sigma_0(T^4 - T_0^4) + S\eta(T - T_0) + Q_T. \tag{6.15}$$

Here S is the total area of the sample surface, σ_{SB} is the Stefan–Boltzmann constant, σ_0 is the emissivity and η is the constant of convective heat exchange.

While the convection losses and the losses by thermal conduction through thermocouples depend only on T, the radiative losses are determined by the condition of the target surface (through the factor σ_0). This is why, under conditions of oxidation, when the thickness of the oxide layer differs for the same temperature during laser heating and cooling of the sample, a difference between the values $\sigma_0^{+}(T_i)$ and $\sigma_0^{-}(T_i)$ would be expected to occur. Nevertheless the calculations show that, with moderate heating temperatures, the condition $(A_R - A)/A \ll 1$ is always fulfilled. For example, even with relatively large changes in the degree of blackening $\sigma_0^+ - \sigma_0^- = 0.2$ (close to the upper limit of this variation), for $P = 20$ W, $T = 1000$ K, $S = 0.3$ cm^2 and a characteristic value of absorptivity with such target temperatures, $A \sim (0.2\text{–}0.5)$, one gets $(A_R - A)/A = (3.5\text{–}8.5) \times 10^{-2}$. Note that the difference between σ_0^+ and σ_0^- could be larger at lower temperatures, but in this case the radiative losses are insignificant.

It is evident that in order to increase the accuracy of the determination of $A(T)$ [211], the average power of the radiation has to be increased, and also a sufficiently rapid recording of the curves $T(t)$ and $\mathrm{d}T(t)/\mathrm{d}t$ should be ensured.

In order to establish the necessary conditions for applying this calorimetric method, as well as for analysing the results obtained by its application, it is essential to assess the feasibility of modelling the foil as thermally thin, or more specifically the drop in temperature between the irradiated, and the opposite, faces of the target should be sufficiently low [213]. With uniform illumination, the drop in temperature between the irradiated and the opposite faces of the sample is $\Delta T = PAh/(S_s k_T)$, where S_s is the irradiation area. For example, in the case of a copper sample of thickness $h = 0.1$ cm for $S_s = 0.1$ cm^2, $P = 20$ W, $k_T = 3.2$ W cm^{-1} K^{-1} and $A \sim 0.2\text{–}0.5$ one finds $\Delta T \approx 1\text{–}3$ K, i.e. much less than the values obtained with the oxidation activation regime, $T \leqslant 1000$ K. It is evident that when shifting to metals showing a lower conductivity, as for example iron and its alloys, titanium and others, ΔT will increase. Nevertheless even in these cases a suitable selection of experimental parameters may ensure the necessary conditions that $\Delta T \leqslant 10$ K.

A typical experimental CRO trace [208] of the evolution of signals $T(t)$ and $\mathrm{d}T(t)/\mathrm{d}t$ in the process of heating a copper sample by CW CO_2 laser irradiation, as well as during its subsequent cooling, is presented in figure 6.2 (in order to simplify data processing the polarisation of the $\mathrm{d}T/\mathrm{d}t$ signal was changed at the laser switch-off moment, t_f). The calibration of the $T(t)$ signals was made by help of certain, familiar means (related, for example, to the temperature of certain phase transitions). The signals $\mathrm{d}T/\mathrm{d}t$ were normally calibrated by the processing of 'cool' heating curves (up to 50–100 °C). The influence of thermal losses and the $A(T)$ dependence can be neglected in this case, and the sample temperature increases according to a linear law

$$\frac{\mathrm{d}T}{\mathrm{d}t} = \frac{AP}{cm}. \tag{6.16}$$

Then A can be estimated for known values of P, m, c, from the slope of the $T(t)$ curve. This magnitude has a corresponding value of the initial step on the $\mathrm{d}T/\mathrm{d}t$ curve, which enables one to make the necessary calibration when selecting targets with different $A(T_0)$ and m values.

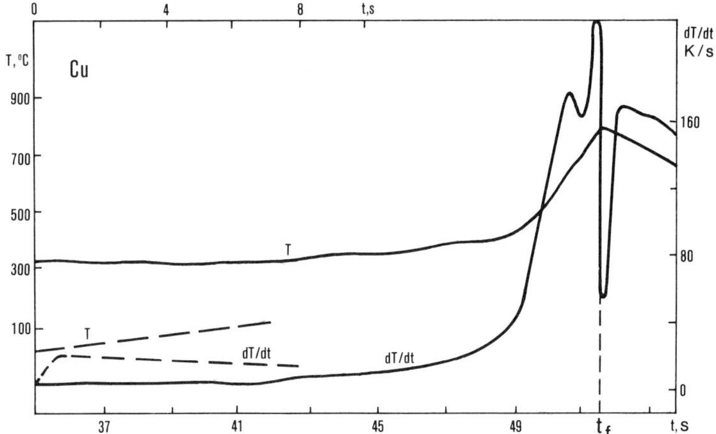

Figure 6.2 Typical CRO traces of $T(t)$ and $dT(T)$ signals recorded during air-heating of a copper foil under the action of CW CO_2 laser radiation, and its subsequent cooling after switching off the laser source (with $t = t_f$). The broken curve indicates the initial regions of the heating curves (the time scale is specified on top), $P = 21$ W, $A_0 = 0.02$, $m = 50$ mg.

When the condition $AP \gg P_{ex}$ is no longer valid, the effective absorptivity $A_e(A_e > A)$ can be approximated by

$$A_e(T_i) = A + \frac{P_{ex}}{P} = \frac{cm}{P}\frac{dT}{dt}(T_i)\bigg|_{-} + \frac{Q_-(T_i)}{P} \tag{6.17}$$

where the power of thermal losses, $Q_-(T_i)$, is evaluated by extrapolating the curve $dT/dt|_-$ recorded at relatively low target temperatures (when the thermal release of the chemical reaction is insignificant within the temperature range considered as being of interest).

Let us mention another variant [209] of the calorimetric method for the determination of absorptivity, which is adequate in investigating laser radiation effects on certain thin blades (of thickness h) with transverse dimensions, L, considerably exceeding the $2R_s$ diameter of the irradiation spot. Under these circumstances, with conditions $h^2 \ll \kappa_M \tau_p$, $L^2 \gg \kappa_M \tau_p$, $R_s^2 \ll \kappa_M \tau_p$ fulfilled, and neglecting all types of losses from the irradiation area with the exception of radial thermal dissipation on the metal blade, the absorptivity results from the equation

$$\frac{d(\Delta T)}{d(\ln t)} = \frac{AP}{4\pi k_T h}. \tag{6.18}$$

This method enables the estimation of only the 'cold' values of target absorptivity (including those obtained after metal oxidation under the action of laser radiation), which strongly limits its efficiency in the study of dynamic aspects of heating.

Together with the estimation of the variation of absorptivity during the heating of targets under the action of radiation, additional information is also supplied by the determination of the reflectivity of the oxidising target. The simultaneous determination of A and $1 - R$ values enables, first, the direct estimation of the thermal power, P_{ex}, released by the exothermal oxidation reaction [214]. On the other hand, the recording of the variation of R offers the chance for additional control of the accuracy in the evaluation of absorptivity by a calorimetric method. In the third place, the combination of a pyrometric method for the determination of the irradiated surface temperature with recordings of reflectivity variation allows for the study of heating dynamics with metal targets of any shape and size [215]. The main difficulty consists, in this case, in the low accuracy of the estimation of reflectivity estimation during the initial heating stages (until oxidation activation) when, for metals (and especially corresponding to an incident radiation wavelength of $\lambda \simeq 10.6\ \mu m$) R is close to unity.

The recording of the specularly reflected component of the light flux is not sufficient for establishing the exact values of reflectivity, as the radiation fraction scattered at the oxidising metal surface may become significant. Besides, the intense heating under the action of laser radiation induces appreciable thermal deformation on the metal surface leading to alterations in the direction of the reflected radiation. For this reason photometric (integrating) spheres are used in order to correctly evaluate the sample reflectivity R.

An experimental set-up [214] enabling simultaneous determination of A and R is shown in figure 6.3. Continuous wave CO_2 laser radiation ($P \simeq 20$ W) was focused through a lens on the surface of the metallic foil placed in air, close to the slit in the wall of the photometric sphere. The angle between the laser beam and the normal to the target surface was selected in the range 10–20°, while the diameter of the sphere was $\simeq 100$ mm. A laminated aluminium foil of high reflectivity, ensuring an adequate scattering for the $\lambda \simeq 10.6\ \mu m$ radiation, was used to cover the internal surface of the sphere [215]. The reflectivity of the irradiated sample was recorded using sensitive photodetectors whose signals were passed through a lock-in amplifier and subsequently introduced into one of the oscilloscope channels, while the other three channels were used for controlling the time evolution of the radiation power, $P(t)$ and the signals, $T(t)$ and $\mathrm{d}T(t)/\mathrm{d}t$, involved in the evaluation of $A(t)$. Light emission by the heated target itself was eliminated by the mechanical modulation of the incident laser light at a frequency of 550 Hz and with the help of interfering filters (having a transmission peak in the range $\lambda = 10.6\ \mu m$ and a transmission bandwidth $\Delta\lambda = 0.45\ \mu m$), placed in front of the detector.

The blowing of a gas jet (air, helium, nitrogen, etc) at a certain moment

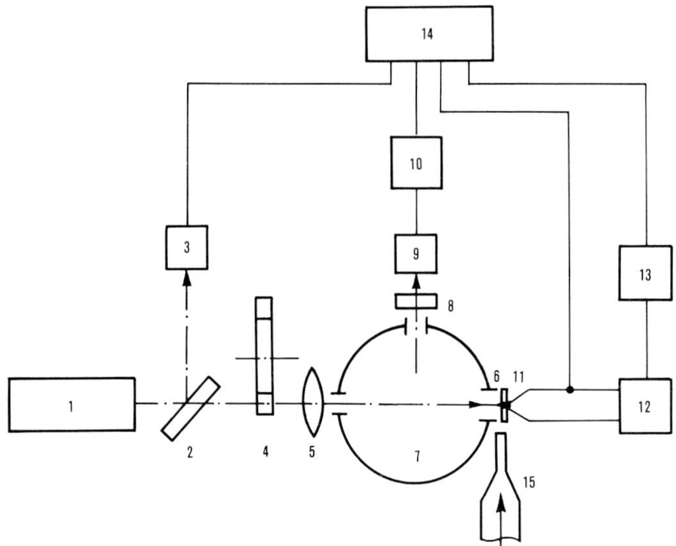

Figure 6.3 An experimental set-up designed for simultaneous determination of the effective absorptivity and of the metallic foil reflectivity. (1) cw CO_2 laser source; (2) beam-splitter; (3) power meter; (4) mechanical modulator; (5) optical lens; (6) metallic sample; (7) photometric sphere; (8) dispersive filter; (9) pyroelectric detector; (10) selective amplifier; (11) thermocouples; (12) *RC* differentiator; (13) amplifier; (14) oscilloscope; (15) gas blower.

during the interaction process may be applied in order to efficiently influence the kinetics of heating and oxidation. From this viewpoint, the experimental arrangement presented in figure 6.3 has additional advantage. The shift of the irradiated foil towards the extremities of the photometric (integrating) sphere enables the investigation of the optical characteristics of the sample concurrently with blowing a gas jet on the sample, thus modelling, to a certain extent, the conditions characteristic to gas jet laser cutting of metals.

6.4. The low-temperature phase in the oxidation of metals and alloys

Let us consider the thermodynamic processes occurring at the surface of metal samples irradiated until the moment of activation of the oxidation reaction, i.e. the low-temperature oxidation phase.

6.4.1. The growth of thin oxide films

Besides the laser cleaning effect (and in competition with it), low-temperature heating determines, for the majority of metals, the formation of thin oxide layers. The study of these layers for the irradiation in air of copper targets was reported in reference [216].

Polycrystalline copper foils 0.3 mm thick, carefully polished using a diamond paste, were subject to irradiation with a cw CO_2 laser. Owing to the relatively high thermal losses due to the transverse dimensions of the sample ($5 \times 5\,\text{mm}^2$), at an incident laser power of $\simeq 20$ W the target temperature relatively quickly (i.e. in a time interval $t = t_{st} \simeq 1$ s) reached a stationary value $T_{st} = (220 \pm 5)\,°\text{C}$. For this reason, at $t \gg t_{st}$ the laser oxidation of samples showed an isothermal character.

A picture of a copper foil surface, laser oxidised in air for 660 s is presented in figure 6.4. Superficial scratches resulting from polishing may be seen. The oxide crystals are relatively small and have characteristic sizes of $\simeq 0.05\,\mu$m. The oxide layer is highly adherent to the metal blade. It is interesting to mention that for irradiation durations of $t_f \leqslant 2000$ s the sample absorptivity was not modified (within the limits of measuring errors), keeping the same value both before, A_0, and after, A_f, laser irradiation, i.e. $A_0 \simeq A_f \simeq (1.25 \times 10^{-2}–1.3 \times 10^{-2})$.

Investigation of the thickness and contents of thin oxide layers [216] was carried out according to the following method of sample preparation: a protective screen film was deposited on the oxide layer and the sample was polished in a special solution jet (D_2–Struers) starting with the non-irradiated face of the sample. Jet concentration in a certain area located in the middle of the foil resulted in a perforation with very sharp edges. In an area where the adhesion of the oxide layer to the metal was poor, it was possible to detach it from the foil. A magnified image of this oxide film obtained for an irradiation time $t_f = 1900$ s is presented in figure 6.5, the picture being rotated so as to emphasise the edge.

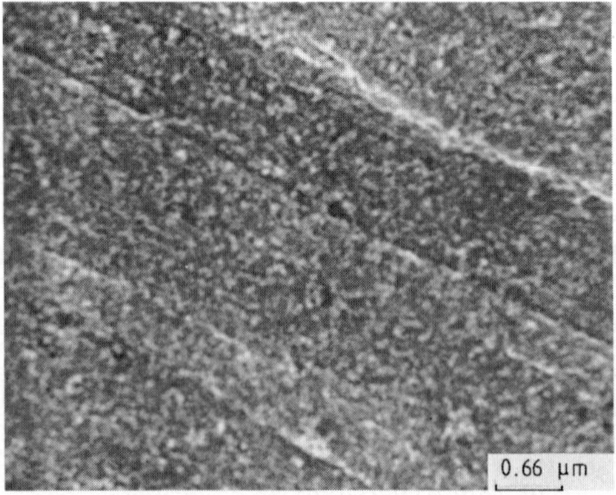

Figure 6.4 Scanning electron microscopy (SEM) image of the irradiated region on the surface of a copper target; irradiation time $\simeq 660$ s; $T \simeq 215\,°\text{C}$.

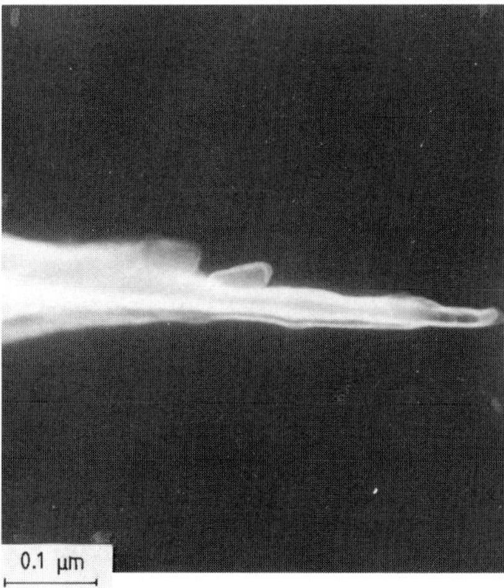

Figure 6.5 SEM image of the edge of a self-sustained oxide layer obtained by thinning a laser-oxidised copper sample. The sample was irradiated for $\simeq 1980$ s at $T = 220\,^{\circ}\mathrm{C}$.

The thickness of the oxide layer can be evaluated on this basis as being $x = 500$ Å. Transmission electron microscopy (TEM) studies were conducted after again rotating the film into such a position as to enable a normal incidence of the electron beam on its surface. Micrographs of the oxide layer as well as the corresponding electron diffraction pattern are presented in figure 6.6. Processing of such photographs, corresponding to different regions on the sample surface, showed that the film formed during initial stages of laser oxidation of copper consists of a single oxide type, namely Cu_2O. This explains the observed identity, $A_0 \simeq A_f$. Indeed, in case of Cu_2O, $n \approx 1.5$–2.5 [217] and for $x = 0.05\ \mu$m, the application of equation (6.7) leads to $(A_f - A_0)/A_0 \leqslant 4.6 \times 10^{-3}$.

From the data obtained in [216] it results that even prolonged ($t_f \leqslant 40$ min) laser oxidation of a polished copper foil at low temperatures ($T_{st} \simeq 220\,^{\circ}\mathrm{C}$) would not determine changes in its absorptivity. This result is of particular importance for those interested in the study of the metal optics of CO_2 lasers, with which copper mirrors are commonly used as a reflecting medium.

Among the various methods of increasing the corrosion resistance of metallic mirrors, the implantation of different ions in the metal support has recently received much attention. One of the advantages of ion implantation is the absence of a sharp boundary between the support and the surface layer, in contrast with the coatings made by vacuum deposition.† The resistance

† Occurrence of such a clear-cut boundary may lead to a worsening of the strength and thermal conductivity characteristics of the mirror and, subsequently, to a decrease in its damage threshold.

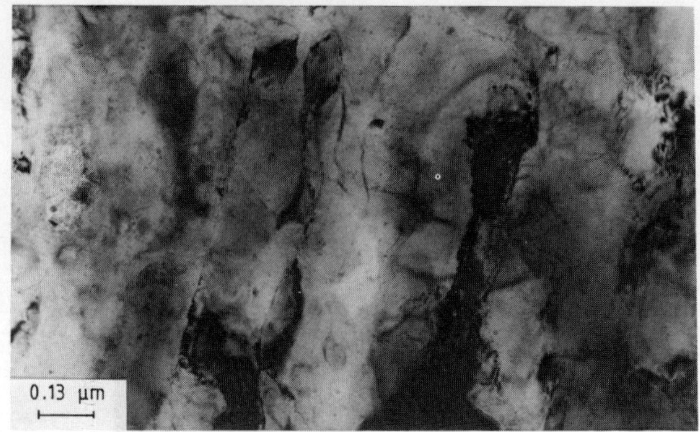

Figure 6.6 SEM images obtained through transmission electron microscopy and electron diffraction, respectively, of the oxide layer peak in figure 2.5, which was placed in a position such that the electron beam was perpendicular on its surface.

to damage of copper mirrors implanted with aluminium ions under the action of radiation pulses of $\lambda = 1.06\,\mu$m was studied in reference [218]. It was shown that the oxidation reaction is strongly slowed down as an effect of the formation of a superficial protective layer of $CuAlO_2$.

Oxidation of ion-implanted copper mirrors was studied in much more detail in reference [219]. The targets consisted of electrochemically polished copper foils implanted with Ag^+ ions (at 80 keV energy, 10^{16} cm^{-2} dose—close to the limiting current density of 0.4 μA cm^{-2}). The targets were oxidised in air at $T_{st} \simeq 200$ °C under the action of cw CO_2 laser irradiation lasting for 40 min. The distribution of elements within the surface layers of the sample was studied using a photoelectronic x-ray spectroscopy method (for details see for example [220]).

The distribution of the concentrations of the elements in the case of an implanted, unoxidised sample (the lines Cu $2p_{3/2}$, Ag $3d_{5/2}$, O 1s), can be examined in figure 6.7. It can be noted that the peak of the Ag^+ distribution is located at a depth of $\simeq 40$ Å.

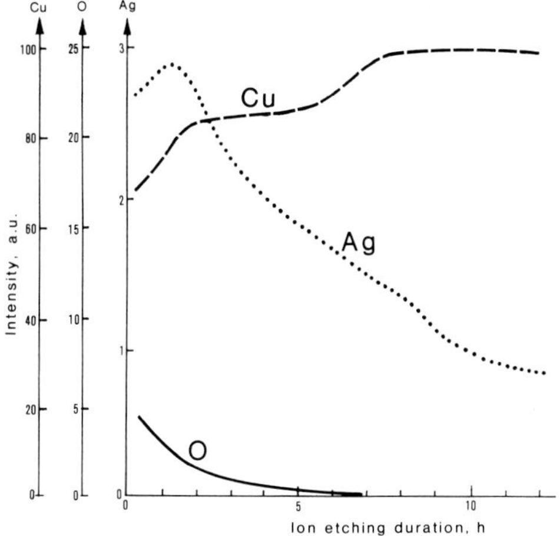

Figure 6.7 Cu, O and Ag concentration profiles, in the case of an implanted, non-oxidised copper sample.

The profiles of Cu and O for an oxidised but not implanted copper sample are presented in figure 6.8(*a*). The value of the electron binding energy for the level Cu $2p_{3/2}$ on the sample surface, $E_b = 932.1$ eV, corresponds to the oxidation phase Cu_2O. An increase in the binding energy which approaches $E_b = 932.4$ eV (pure copper) is noted with the increase in sample depth. The analysis of the position of the oxygen 1s line indicated two peaks (at $E_b = 530.1$ eV and $E_b = 532.1$ eV, respectively). The first corresponds to the occurrence of Cu_2O while the second to the oxygen which is adsorbed on the surface and which disappears after only a few minutes of ionic polishing (its thickness does not exceed a few monolayers). We must note that the thickness of the oxide layer as evaluated during ionic polishing [219] agrees with the results obtained by electron microscopy investigations of the oxide films formed on copper foils under similar conditions of laser irradiation, as mentioned above [216].

The profiles of the concentration of the elements in the case of oxidised copper targets originally implanted with Ag^+ ions are given in figure 6.8(*b*). It can be seen that in this case the thickness of the oxide layer is nearly half that of unimplanted oxidised samples. More precisely, the profile of the oxygen concentration is sharply interrupted in the proximity of the silver concentration peak, which confirms the protective role of the implanted layer. The strong deformation of the Ag^+ concentration profile as compared to the initial one (as can be seen from the comparison of figure 6.7 and 6.8(*b*)) has also to be mentioned.

The improvement in the corrosion resistance of copper mirrors by ionic implantation does not result, as has been mentioned [219], in a significant increase in their absorptivity. For example, the initial value of sample

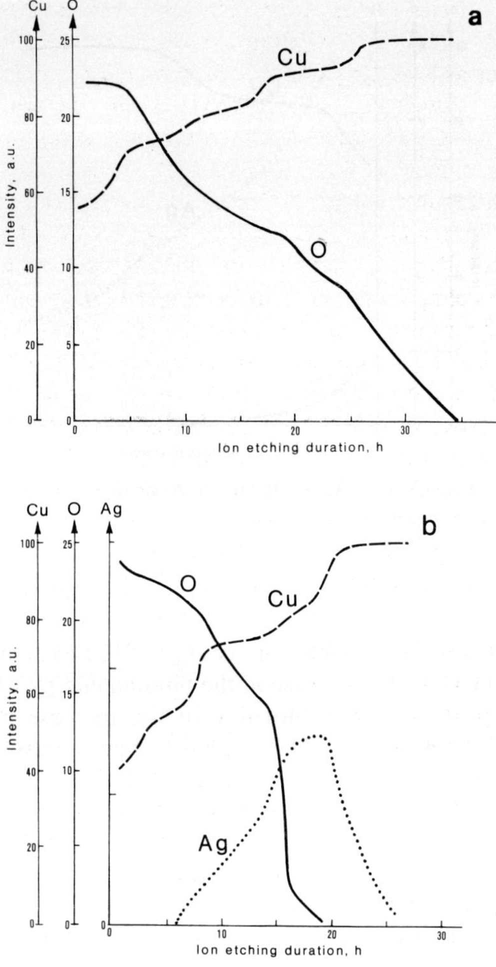

Figure 6.8 (*a*) Cu and O concentration profiles for a copper sample, non-implanted, oxidised for $\simeq 40$ min at $T = 200\,^{\circ}\mathrm{C}$. (*b*) Cu, O and Ag concentration profiles for a copper target, implanted, oxidised for $\simeq 40$ min at $T = 200\,^{\circ}\mathrm{C}$.

absorptivity (A_0) ranged between 0.0085–0.0095, whereas a certain increase, up to $A \simeq 0.015$–0.0125, was noted after implantation.

After laser oxidation the absorptivity of the samples levelled out at a lower level of $A_f \simeq 0.011$, a fact that can be attributed to the annealing of the radiative defects in the superficial layer of the sample. In fact, pulsed laser annealing is expected to result in even more significant improvements of the optical properties of ion-implanted mirrors, preserving their high corrosion resistance.

The growth kinetics of thin oxide films was not studied in the above-mentioned work [216, 219]. This topic was first studied [221] using a neodymium laser as a heating source, which generated pulses of 50 ns–1 ms. Chromium films of thickness 1000–3000 Å were vacuum deposited on quartz substrates. Irradiation was performed in air, monitoring the evolution of the thickness of the oxide film formed on the chromium layer as an effect of heating under the action of laser radiation, by using a quartz resonator [222] (highly suitable and sufficiently simple for this type of study).

The experimentally recorded evolution of the thickness x during the action of the laser pulses of $\tau_p \simeq 1$ ms, at an incident intensity of $I = 6 \times 10^2$ W cm^{-2}, is presented in figure 6.9. It can be noted that the thickness of the resulting Cr_2O_3 layer is not high as it does not exceed 60 Å, which agrees with the values obtained by the numerical solution of the thermochemical problem using the Cabrerra–Mott equation (6.1) and the corresponding thermal conductivity equation for constant absorptivity.

The growth kinetics of Cu_2O films formed on copper by isothermal oxidation in air at low temperature under the action of cw CO_2 laser radiation was studied in [223]. The time, t_f, and temperature, T_{st}, at which oxidation was performed as well as the measured values of the film thickness are presented in table 6.1.

Together with the experimental data, the calculated values of the oxide layer thickness x_1 and x_2 are listed in the table. The x_1 values were obtained by the solution of the parabolic equation (6.2) under isothermal conditions. The values $d_0 = 1.4$ cm^2 s^{-1} and $T_d = 15\,500$ K, established by laser experiments [224] and corresponding to the high-temperature heating stage, were used as thermal diffusion constants (see Chapter 7). For this reason, the fact that x and x_1 get close to each other, as can be noted in this case, has to be regarded as a coincidence. The approximation of the experimental data within the range $x \sim 1000$ Å, corresponding to $T \simeq 150$–250 °C, by the cubic law $dx/dt = (d_0/x^2)\exp(-T_d/T)$, which, according to [225], can be

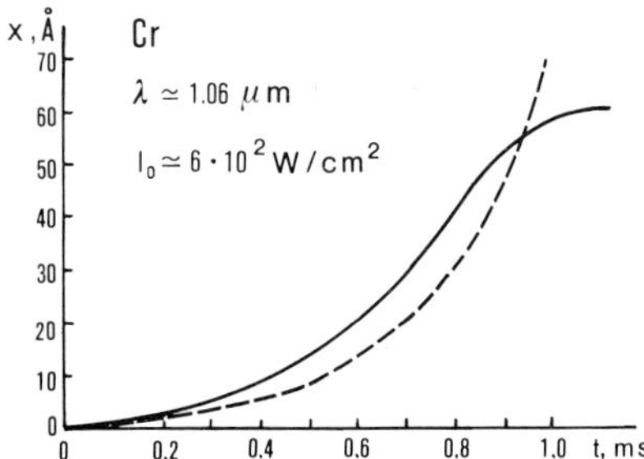

Figure 6.9 Oxidation kinetics of a thin chromium layer ($\simeq 2000$ Å); $\lambda \simeq 1.06$ μm, $I_0 = 6 \times 10^2$ W cm^{-2}; full curve, experimental data; broken curve, numerical calculations.

Table 6.1 Experimental and theoretical values of the thickness of the oxide layer grown on the surface of a copper sample as a result of cw CO_2 laser irradiation in air [223]. MP, mechanical polishing; ECP, electrochemical polishing.

No	Surface preparation before irradiation	t_f (s)	T_{st} (°C)	x (Å)	x_1 (Å)	x_2 (Å)
1	MP	600	155	400–600	285	60
2	ECP	660	195	500–680	660	275
3	MP	1140	210	600–800	1050	610
4	MP	1140	210	600–1000	1050	610
5	ECP	1200	228	800–1000	1465	1110
6	MP	1440	220	1000–1300	1360	945
7	MP	1540	260	1200–1800	1870	1885

applied within this laser oxidation range, appears to be correct. Using the constants $d_0 = 10^{-7}$ cm^2 s^{-1} and $T_d = 12\,775$ K [223], the x_2 values agree with the experimental ones.

Finally we point to the fact that, at the present stage, we have insufficient knowledge to develop theoretical models adequate for the understanding of the growth of thin oxide films subject to laser irradiation, even under conditions of isothermal oxidation. The number of experimental data are still insufficient, and this is the reason why the similarity between the kinetic characteristics of laser irradiation and common non-laser heating (in a thermal reactor, for example) still has no satisfactory explanation.

6.4.2. Oxygen diffusion into the metal foil

The optical properties of an important number of investigated metals (not only Cu which was discussed above, but also Al [211], Fe [209, 211, 215], W [215, 226], V [227–229] and others) are not directly determined by the thermochemical processes in case of laser oxidation stages at low temperatures. Nevertheless the situation is considerably changed with the heating of Ti targets (and possibly with other transition metals too).

It was established [230, 231] that during the initial stage of laser heating of titanium targets in air, a marked increase (by 1.5–2 times) in target absorptivity occurs, preceding the occurrence of the oxidation reaction (which is accompanied by a strong growth in the oxide layer). A typical evolution of the temperature dependence of Ti absorptivity, $A(T)$, under irradiation with cw CO_2 lasers is shown in figure 6.10. It can be noted that heating the sample from room temperature up to $T \simeq 700$ °C (in 3 s) results in a rapid increase in absorptivity until a stationary value A_{st} is reached, which is maintained until the oxidation reaction takes place. Moreover, such evolutions may require considerably shorter periods of time [230], and are irreversible in character. The dependence of A/A_0, for a 50 μm thick titanium

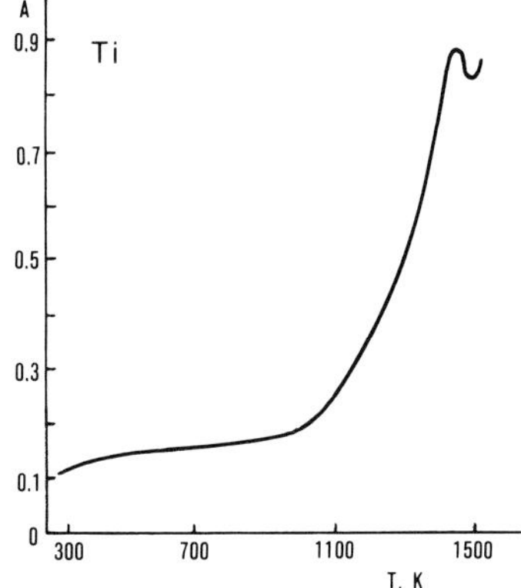

Figure 6.10 Experimentally determined dependence on temperature of the absorptivity of a titanium foil.

foil, on the energy density, E_s, incident onto its surface in a millisecond pulse generated by a neodymium–glass laser source ($\lambda = 1.06\ \mu$m), which results in the foil heating up to a relatively low temperature ($T \leqslant 800\ ^\circ$C), is presented in figure 6.11.

The well known process of oxygen dissolution into metal was considered as an explanation of this behaviour [202, 230–232]. For example, when certain metals such as titanium, zirconium, hafnium and others are heated in oxygen or air, oxygen diffusion and dissolution into the metal support is intensified along with the growth of the oxide film. Consequently, oxygen enrichment of the metal surface layer occurs, which results in the increase of both the electron diffusion on impurities and of the electron mobility within the metal. A significant increase in the sample absorptivity without additional dissipation of laser energy into the oxide layer is noted as a result.

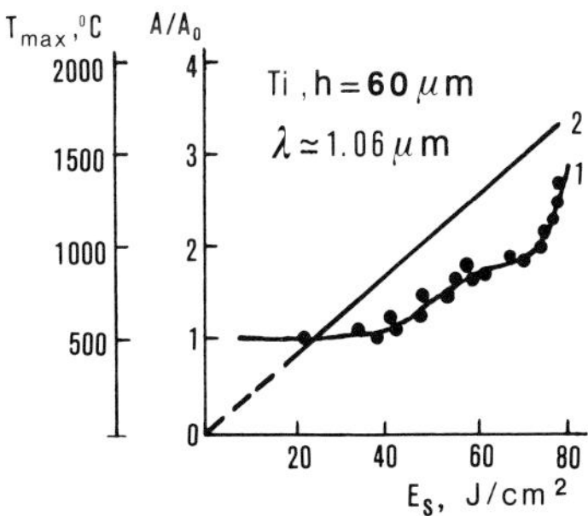

Figure 6.11 Irreversible evolution of the relative absorptivity A/A_0, at the wavelength $\lambda = 10.6\ \mu$m (curve 1), and the dependence of the maximum temperature T_{max} (curve 2) of a titanium foil having a thickness $\simeq 50\ \mu$m, as a function of the neodymium pulsed laser energy density incident onto its surface ($\lambda = 1.06\ \mu$m).

A theoretical model enabling one to take into consideration oxygen dissolution into the metal has been proposed [232]. In the absence of temperature gradients, the diffusion process is described by the equation

$$\frac{\partial N}{\partial t} = D(T)\frac{\partial^2 N}{\partial z^2} \text{ for } z=0 \qquad D(T) = D_0 \exp\left(-\frac{T_D}{T}\right) \tag{6.19}$$

where $N = N(z, t)$ is the oxygen concentration in the proximity of the metal surface (z is the in-depth coordinate of the irradiated target), and T_D is the diffusion activation temperature. Equation (6.19) has to be complemented with the initial and boundary conditions

$$N(z,t)\Big|_{t=0}^{z>0} = 0 \qquad N(z,t)\Big|_{z=0} = N_0 \qquad N(z,t)\Big|_{z=\infty} = 0. \tag{6.20}$$

Within the normal skin effect approximation which defines the shape of the $A(N)$ function, the following formulae are obtained from (6.19) and (6.20)

$$A(t) = \begin{cases} A_{01} + (A_{02} - A_0)y & 0 \leqslant y \leqslant \dfrac{\sqrt{\pi}-1}{\sqrt{\pi}} \\ A_{02} - \dfrac{A_{02} - A_{01}}{\pi[y - (2-\sqrt{\pi})\sqrt{\pi}]} & y > \dfrac{\sqrt{\pi}-1}{\pi} \end{cases} \tag{6.21}$$

where $y = \delta^{-1}\sqrt{\theta/\pi}$, and the function $\theta = D_0 \int_0^t \exp(-T_D/T(\tau))\,d\tau$. The characteristic values of absorptivity A_{01} and A_{02} are given by the expressions

$$A_{01} = \frac{c_0}{\pi\sigma_0\delta_0} \qquad \frac{A_{02} - A_{01}}{A_{01}} = \frac{N_0 v_F \sigma_{ed}}{\Gamma_{ep}}. \tag{6.22}$$

Here σ_0 is the conductivity of the pure metal, δ_0 the metal skin layer depth (see equation (1.45)), Γ_{ep} the frequency of electron–phonon collisions, ω the circular frequency of radiation, v_F the velocity of the conduction electrons on the Fermi surface and σ_{ed} the cross section of electron scattering on impurities.

In order to perform numerical calculations in practical cases when laser heating of metals occurs under oxygen saturation, equation (6.21) has to be matched with a corresponding thermal conductivity equation. Such attempts were conducted in [232]. Nevertheless their value has to be regarded rather from the viewpoint of methodology, as illustrating the possibility of describing qualitatively the results obtained in the experiments with thermally thin and thermally insulated titanium targets.

A more complete quantitative approach is much more difficult to perform, due to uncertainties in selecting the constants A_{01}, A_{02}, D_0 and T_D. For this reason new detailed investigations are in order, in view of clarifying whether or not the sample absorptivity may increase as a result of saturation of the metal with oxygen.

In connection with the difficulty of dealing with this effect, we mention the results obtained in reference [233], where the structure and content of the surface layer of laser-irradiated titanium samples were investigated by x-ray methods. The typical phase content of this surface layer is presented in figure 6.12. It can be noted that the interaction zone consists of an oxide mixture and that there exist no clear boundaries between phases. It is also necessary to mention that the experimental method applied in [233] did not allow one to make a distinction between the titanium oxides TiO and $TiO_{p/q}$ ($p/q < 1$, p and q are integers) and the dissolution of oxygen into titanium. Also, there was no way of specifying the position of the metal–surface interface. For these reasons, it is possible to consider the observed increase in absorptivity as resulting not only from oxygen dissolution into the metal, but also from the formation on its surface of a sufficiently thick oxide layer of TiO, resembling the metal [234] from the point of view of its optical properties in the infrared spectral region.

6.4.3. *Changes occurring in the content of alloys*

An interesting effect occurring in the laser heating of alloys has been observed [235]. The effect consists in a change in the relative concentration of one or more of the alloy components in the surface layer, as a consequence of the fact that some of them undergo stronger oxidation then others. For this reason the change in sample absorptivity noted at the start of laser irradiation does not occur at the expense of additional energy dissipation within the oxide film, but it results from changes occurring in the content and optical properties of the surface layer of the alloy.

In the afore-mentioned experiments [235], a steel foil 100 μm thick was subjected to irradiation in air with pulses from a neodymium–glass laser, operating in free-running mode. The radius of the irradiation area ($\simeq 4$ mm)

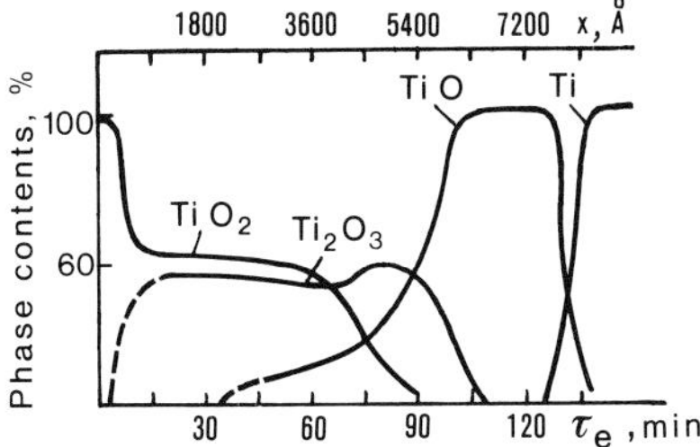

Figure 6.12 Phase distribution in the oxide layer depth in relation to the air–oxide interface, in the case of a laser-oxidised titanium sample (τ_e, etching time).

considerably exceeded the sample thickness, h, and the effective duration of thermal action ($T > T_0$) was ~ 50 ms. Sample absorptivity prior to (A_0) and after (A_f) laser oxidation was determined by using a probing cw CO_2 laser. The ratio A_f/A_0 as a function of temperature, T, reached on the sample surface by the end of the heating action of the laser pulse is presented in figure 6.13.

It can be noted that the ratio A_f/A_0 decreases by nearly a factor of two with increase in temperature, T, up to a value $T_{min} \simeq 400$ °C, i.e. until the initiation of a significant increase in oxide absorptivity. The measurements reported in reference [235] for characterising the spatial distribution of different phases, showed that (figure 6.13(b)) while iron and chromium oxides are present on the surface of the irradiated samples, nickel oxides are absent. As a result, nickel concentration within the surface layer of the alloy increases by 30% compared with the nickel concentration in the bulk of the sample.

Analysis of the surface layer of irradiated samples has allowed a qualitative estimation of the decrease occurring in the evolution of the ratio A_f/A_0 with T, for $T \leqslant T_{min}$. For example, within the normal skin effect approximation one has $A_M(T) \sim \sqrt{r_M(T)}$, where $r_M(T)$ is the electrical resistivity. According to available data [236], r_M decreases with increase in the nickel concentration and the corresponding decrease of the chromium concentration in the alloy Fe–Ni–Cr. Consequently, during laser oxidation of the alloy one has $A = A_M$ and $A_f/A_0 \sim \sqrt{r_f/r_0}$ (r_0 and r_f being the values of the initial and final electric resistivity of the alloy, respectively). The numerical estimations performed in [236] gave $A_f(T_{min})/A_0 \simeq 0.7$, which agrees with the experimental data.

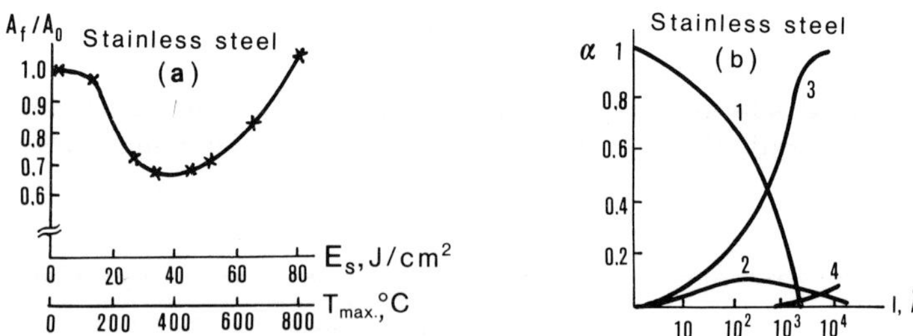

Figure 6.13 (a) Changes in the relative absorptivity, A_f/A_0, of a stainless steel sample, at the wavelength $\lambda = 10.6$ μm, as a result of neodymium pulsed laser irradiation ($\lambda \simeq 1.06$ μm), as a function of the energy density, E_s, incident in air on the surface of the samples. A_0 is the initial value of the absorptivity, T_{max} the pulsed excursion in the steel foil temperature. (b) Distribution of the relative content of the phases in the surface layer of the stainless steel samples exposed to pulsed laser heating up to 700 °C, as a function of the distance l to the oxide–air interface: (1) Fe_3O_4; (2) Cr_2O_3; (3) Fe; (4) Cr.

6.5. Activation of the oxidation reaction

The interdependence between target temperature, T, target heating ratio ($\mathrm{d}T/\mathrm{d}t$), which is determined by the value AP, and the rate of growth of the oxide layer, leads to a rapid increase in the target temperature and absorptivity at $T > T_a$. The temperature T_a and the corresponding irradiation time, t_a, are called activation temperature and activation time of the oxidation respectively (alternatively temperature and time for the development of thermochemical instability).

The characteristic curves $T(t)$ of the temperature evolution of certain identical copper samples heated in air [211] are presented in figure 6.14.† Patterns of $A(t)$ obtained from the processing of $\mathrm{d}T(t)/\mathrm{d}t$ curves (according to the formula (6.13)) are also presented on each diagram.

To be highlighted is, first, the non-monotonic variation of $A(t)$ [211, 237]. In the initial stage of laser heating, i.e. up to $T \leqslant 300$–$500\,^{\circ}\mathrm{C}$, a decrease in absorptivity $A(t)$ with temperature was noted, if the initial sample absorptivity, A_0, exceeded the absorptivity of the pure metal, at room temperature, A_M, i.e. $A_0 \geqslant A_m$. Alternatively, if $A_0 \simeq A_M$ (following careful polishing and alcohol cleaning), the absorptivity remained approximately unchanged, $A(T) \simeq \mathrm{const.}$

For $T > T_a$, the copper sample heating rate exhibits a strong increase simultaneously with the sample absorptivity, which can reach, near the melting point, T_m, a value of $\simeq 0.5$, exceeding by orders of magnitude the value $A_M(T_m)$ which would normally result from the temperature dependence of the pure copper absorptivity [23, 204]. Moreover, at $T > T_a$, the $A(T)$ curves exhibit an oscillatory character (this aspect will be discussed in more detail in the next section).

Considering copper, let us mention the ways in which the irradiation conditions affect the t_a value and, correspondingly, the time necessary for heating the target up to the melting point, t_m. Experimentally, t_a can be measured in various ways. For example, determination of activation time starting from the bending point (if this can be singled out) or from the maximum bending moment of the curves $T(t)$ was proposed in certain works [192, 193]. A simpler and more exact method of measuring t_a as against the minimum of the function $\mathrm{d}T(t)/\mathrm{d}t$ was proposed in [208, 238] for $A_0 > A_M$. Occurrence of the function minimum is due to competition between two factors, i.e. the effect of laser cleaning of the sample surface, compensated, to a certain degree, by the evolution of $A_M(T)$, and the increase in A as a result of the initiation of thermochemical instability.

† It has to be mentioned that until recently copper targets have been the most commonly investigated subject in studies regarding the thermochemical mechanism of CO_2 laser (mostly cw operated) interaction with metal samples.

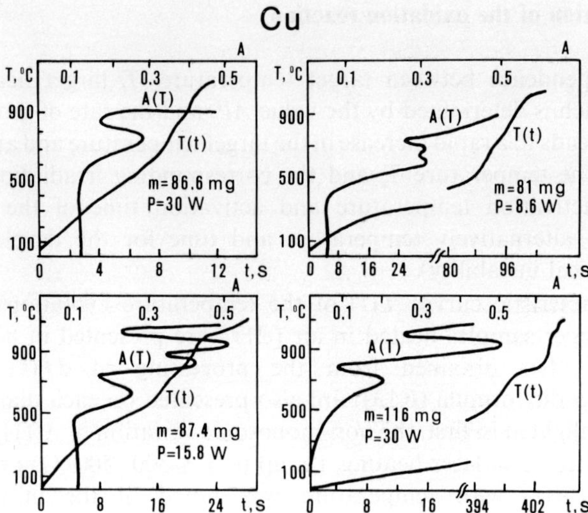

Figure 6.14 Characteristic curves, $T(t)$, of heating in air under the action of cw CO_2 laser radiation of several copper targets, and the respective dependences $A(T)$ for different experimental conditions.

The experimentally established dependences [211] of t_a and t_m on the power of cw CO_2 laser radiation, P, are presented in figure 6.15 for a given value of the initial absorptivity of the copper samples, $A_0 \simeq 0.1$ (the irradiation spot area was $S_s \simeq 0.09$ cm^2). It can be noted that the values t_a and t_m depend non-linearly on the value of the laser radiation intensity, $I_0 = P/S_s$. A similar behaviour was noted [192, 193] with the oxidation of certain chromium layers deposited on massive dielectric samples (modelling the heating of semi-infinite metal targets).

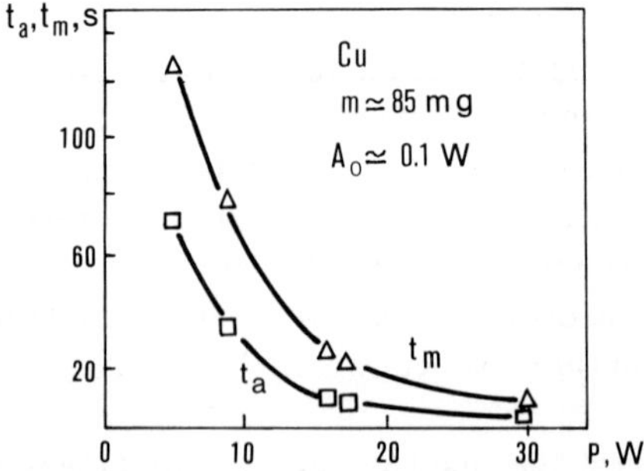

Figure 6.15 Dependence on the incident radiation power of the two characteristic heating times, t_a and t_m, corresponding to oxidation activation and melting, respectively, of several copper foils in air.

The value of the initial absorptivity, A_0, influences the sample heating rate to practically the same extent as the incident laser power, P. To illustrate this fact, the curves $t_a(A_0)$ and $t_m(A_0)$ obtained in the case of laser heating in air, with $P = \text{const}$, are presented in figure 6.16.

These data highlight the special importance attached to the condition of the metal sample surface during laser heating even in a highly oxidising medium (air).

As it was shown experimentally [211], a decrease in the air pressure down to a critical value, p_{cr}, which is $\simeq 0.01$ atm for copper, does not result in a significant change of t_a, i.e. the growth of the oxide layer is not limited by the access of oxygen to the reaction area. A sudden increase in t_a was noted at $p < p_{cr}$ in the experiments conducted with copper targets, while at $p \simeq 3 \times 10^{-4}$ atm it was not possible to melt the targets even with P levels of $\simeq 30$ W [211]. It was also established that the rate of heating of copper samples is determined to a higher extent by the initial absorptivity, A_0, than by the air pressure, p.

The behaviour described above has been confirmed for a large variety of metals. As an example, the characteristic curves $T(t)$ and $A(T)$ for steel and Duralumin targets are presented in figure 6.17. The decrease in $A(t)$ until the moment of oxidation activation as well as the similarity between the shapes of the activation curves obtained with steel and copper targets, are clearly displayed. At the same time the activation of the oxidation reaction does not result in a sensitive increase in the absorptivity $A(T)$, for Duralumin targets heated in the range $T \leqslant T_m$, which would mean that the thermochemical mechanism changing the absorptivity does not act the same way with all metals.

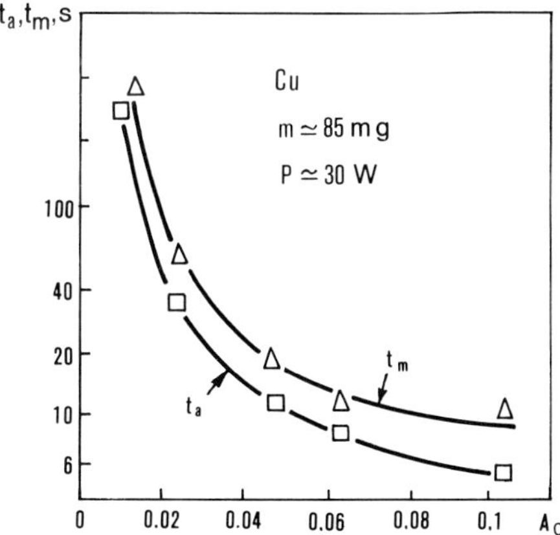

Figure 6.16 Dependence of the activation time, t_a, and of the melting time, t_m, on the initial absorptivity, A_0, of several copper samples.

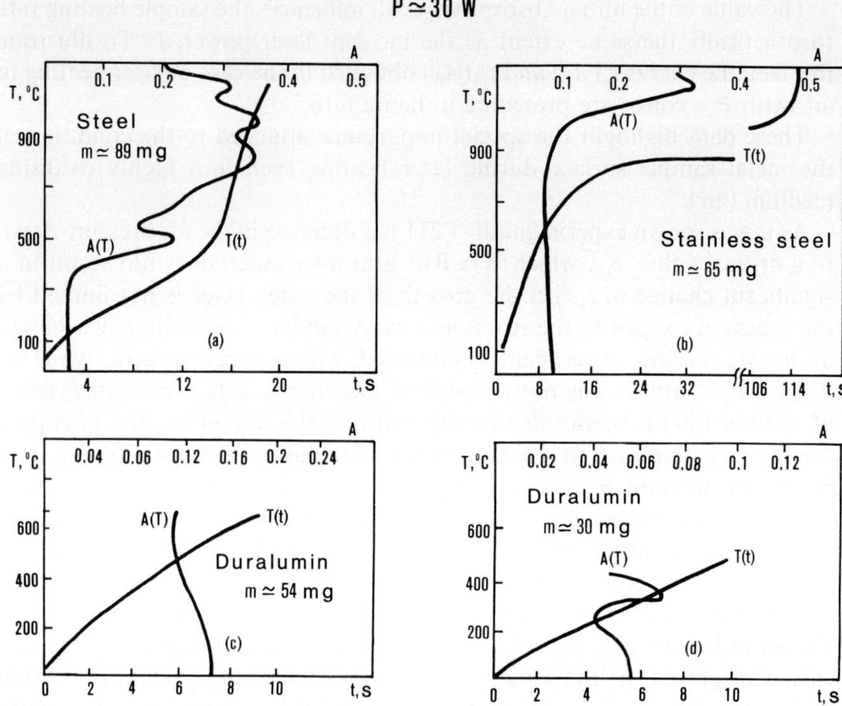

Figure 6.17 Evolutions of the quantities $T(t)$ and $A(t)$ for steel and Duralumin targets exposed to the action of CW CO_2 laser radiation in air ($P = 30$ W).

The main peculiarity with the heating of titanium or zirconium targets consists in the fact that oxidation activation is not preceded by a decreasing, or even constant, evolution of absorptivity $A(T)$; on the contrary, a rapid increase in $A(T)$ (see figure 6.10) is noted in the initial stage of heating. Nevertheless the curves $T(t)$ are slightly different from those characteristic of other metals.

A general approach to the calculation of t_a (one of the main characteristics of the mechanism of oxidation of metals by laser heating) was proposed in [239]. The case when the thermal instability is preceded by a stable evolution of the temperature and the time t_a can be experimentally recorded was also studied. In this case, the value of the stationary temperature, T_a, at which oxidation activation may occur is determined from the equation

$$I_0 A = \mu_n T^n \qquad n > 0. \tag{6.23}$$

The constants n and μ_n depend on the mechanism of heat evacuation from the irradiation area. For example, in the case of heat evacuation by thermal conductivity, $n = 1$ and $\mu_n = k_T/R_s$ (k_T stands here for the metal thermoconductivity and R_s is the radius of the irradiation spot); for the heat losses by radiation, $n = 4$, $\mu_n = \sigma_{SB}\sigma_0$.

Also, in reference [239] the equation of chemical kinetics was given in a generalised form

$$\frac{dx}{dt} = \frac{a_m}{x^m} \exp\left(-\frac{T_d}{T}\right) \tag{6.24}$$

where $m = 0, 1, 2$ correspond, respectively, to linear, parabolic (according to formula (6.2)) and cubic oxidation laws, while a_m is the reaction rate constant. Finally the absorptivity in the thin oxide layers was approximated by the expression

$$A = A_0 + \alpha_k x^k. \tag{6.25}$$

We note that with $k = 2$ formula (6.25) is identical to (6.7).

This formulation of the problem [239] allowed for obtaining the following generalised expression for t_a and $A(t_a)$, of interest for practical applications

$$t_a \simeq \sqrt{\frac{2\pi}{f}} \left(\frac{nf}{k}\right)^f \frac{1}{a_m k} \left(\frac{A_0 T_s}{\alpha_k T_d}\right)^f \exp\left(\frac{T_d}{T}\right) \tag{6.26}$$

where $f = (m+1)/k$. It is to be mentioned that several of the previously established formulae result from the application of formula (6.26) to particular cases. For example, with $m = 1$, $k = 1$, the expression of t_a given in [192, 193] is obtained, while $m = 1$, $k = 2$ leads to the expression mentioned in [238]. In the case of oxidation activation of certain metals, as for example titanium, A_0 has to be replaced with the value of the saturated absorptivity, A_s, in equation (6.26).

The expression of t_a in the case of a non-stationary regime of oxidation activation with a thermally thin metal foil of mass m, was established in [238], taking into account the thermal losses by convection

$$t_a = \frac{cm}{PA_0}\left(\frac{T_d}{\ln(2\alpha_2 a_2 mc/A_0\eta S)} - T_0\right). \tag{6.27}$$

Here η is the constant of convective heat exchange and S the total area of the foil.

Results of calculations based upon the relations (6.26) and (6.27) for A_0 values that do not greatly exceed the A_M value, i.e. when the laser cleaning effect is negligible, are in good agreement with the experimental data.

6.6. Radiation interference in the metal–oxide system

As has been mentioned, the function $A(T)$ may exhibit an oscillating character when $t > t_a$. This effect, noted in [211, 237], was considered [208] as being the result of laser radiation interference within the system of metal–oxide layers.

Interference phenomena in multilayer systems are well known (see, for example, [209]). Nevertheless, their importance for the study of the interaction of laser radiation with materials and especially metals was given attention only after the laser-thermochemistry investigations took off [2, 3, 105, 197]. Even then the possibility of observing certain sharp interference images on the surface of laser-oxidised metal targets (both processed and unprocessed prior to laser oxidation) appeared rather doubtful. Indeed, a prerequisite would have been that the oxide layer growing on the metal exhibits a certain degree of homogeneity and uniform distribution across the irradiation spot. On the other hand, the oscillations of the quantity $A(T, x)$ could have been induced by some other factors (for example the multicomponent structure of the layer as well as the mutual transformation of the oxides).

A series of control experiments was performed on copper targets, while numerical estimations were done in order to check the assumptions above.

First of all it was demonstrated [214] that the oscillations in absorptivity recorded by calorimetric measurements do not result from flaws in the method of investigation. The curves $A(t)$ obtained by the processing of $\mathrm{d}T(t)/\mathrm{d}t$ signals recorded simultaneously with the evolution of the reflectivity $R(t)$ of the copper sample heated in air under the action of a CW CO_2 laser are presented in figure 6.18. It can be seen that over the whole temperature range, $T_0 \leqslant T \leqslant T_m$, the curves $A(t)$ and $1 - R(t)$ are in good agreement.

Secondly [204], copper foils were heated up to a temperature $T > T_a$ and were then exposed to an argon jet in order to preserve the oxide layer thickness obtained during laser heating until that moment and to avoid its further increase as an effect of cooling in air. Transverse microsections of irradiated samples have revealed that the layer thicknesses obtained, $x \simeq 2–10\ \mu m$, are comparable with the laser radiation wavelength, $\lambda = 10.6\ \mu m$, and are situated in the proper range (x^k_{min}, x^k_{max}) for observing interference minima, A^k_{min}, and maxima A^k_{max}, in the absorptivity. For example, for the first minimum and maximum values, one has $(k = 1)$

$$x^1_{min} = \frac{\lambda}{2n} \qquad x^1_{max} = \frac{\lambda}{4n}\frac{z}{\pi} \tag{6.28}$$

where z is the root of the equation

$$\tan z = (1 - n^2)\left(z + \frac{n^2 A_0}{4k}\right). \tag{6.29}$$

Here n and k are the refraction index and the absorption index, respectively, of the laser radiation in the oxide layer.

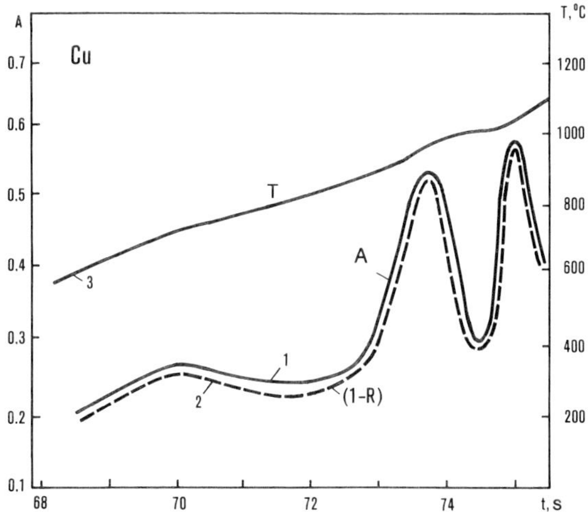

Figure 6.18 Experimental curves corresponding to the time evolution of the absorptivity of a copper sample exposed to cw CO_2 laser irradiation in air: (1) calorimetric determinations; (2) measurements performed by means of a photometric sphere. Curve (3) indicates the temperature variation with time.

The copper oxide layer, with a brown–red colour characteristic of Cu_2O, appeared to be rather homogeneous in thickness. Scanning electron microscopy analysis of oxidised samples [53] has revealed (figure 6.19) that even in the case of the unpolished targets used in [208], for which clear interferential oscillations have been noticed, the characteristic sizes of the inhomogeneities on the oxide layer surface did not exceed 1 μm.

Thirdly, numerical evaluation of the evolution of $A(T)$ [208, 240], performed by solving the equation set (6.2), (6.4), (6.9), using values of Cu_2O

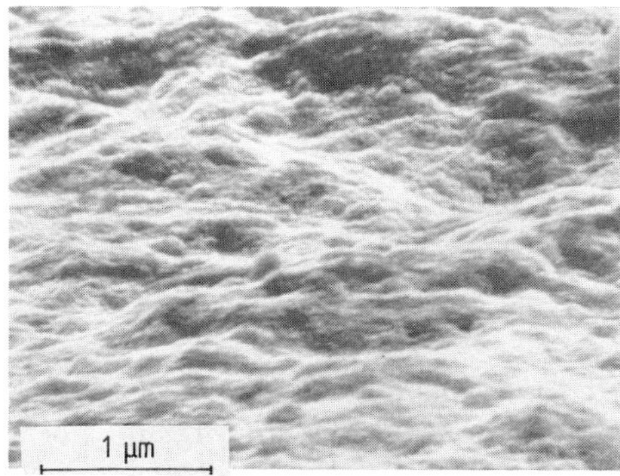

Figure 6.19 Typical SEM image of the surface of unpolished copper targets used in experiments.

optical constants according to the data in [241], have proved to be in good agreement with the experimental curves (figure 6.20).

Fourthly, it has been shown [204] that the dependence on temperature of the oxide's optical constants does not affect the absorptivity of the copper targets during their heating and oxidation under the action of laser radiation. Samples with identical initial absorptivity, A_0, and sizes have been chosen for this purpose. Laser heating was stopped at different values of the final heating temperature, T_f, simultaneously with exposure of the samples to an argon jet. After cooling, the new values of absorptivity of the 'cool' samples, A_0^f, have been determined through a short (~ 1 s), and slight (20–50 °C) laser reheating. The differences observed between the values A_0 and A_0^f have to be ascribed to the laser oxidation process. It is obvious (figure 6.21) that the dependence $A_0^f(T_f)$ is qualitatively analogous to the curve $A(T)$, that is it exhibits an oscillating character. A good quantitative agreement can also be noticed between A_0^f and the absorptivity of the sample, A, which would correspond to a temperature $T = T_f$, that is with $A(T_f)$.

Finally, from the reported experiments [242], it has been shown that the number of oscillations, N, of $A(T)$ is rising when the laser radiation wavelength λ decreases (lasers with $\lambda = 10.6$, 1.06 and 0.515 μm have been used). More exactly, in the case of a weak dependence of the optical constants of the oxide on the laser radiation wavelength one has $N \sim 1/\lambda$. From this angle, we mention the possibility of increasing the sensitivity of the 'interferential' method used to control the growth dynamics of oxide films during the heating of metals under the action of CO_2 laser radiation [243]. Thus we can resort to the recording of target reflectivity by use of probe

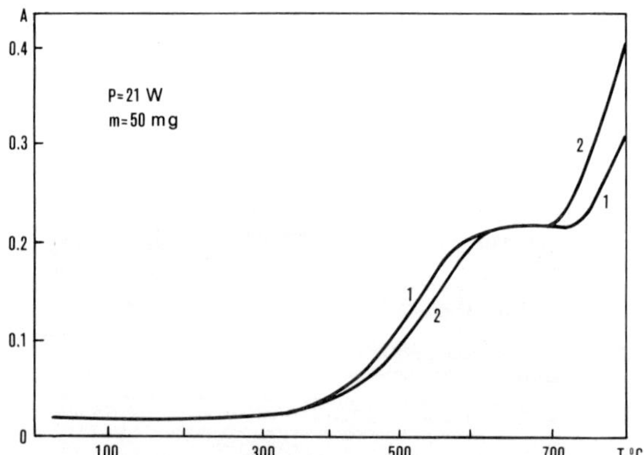

Figure 6.20 Experimentally established evolution (curve 1) of the temperature dependence of the absorptivity, $A(T)$, of the $Cu + Cu_2O$ system, and the corresponding calculated dependence (curve 2) using the following values for the oxide optical and thermophysical constants: $n \simeq 1.1$, $k \simeq 0.019$, $d_0 \simeq 3 \times 10^{-3}$ cm^2 s^{-1} and $T_D \simeq 10\,000$ K.

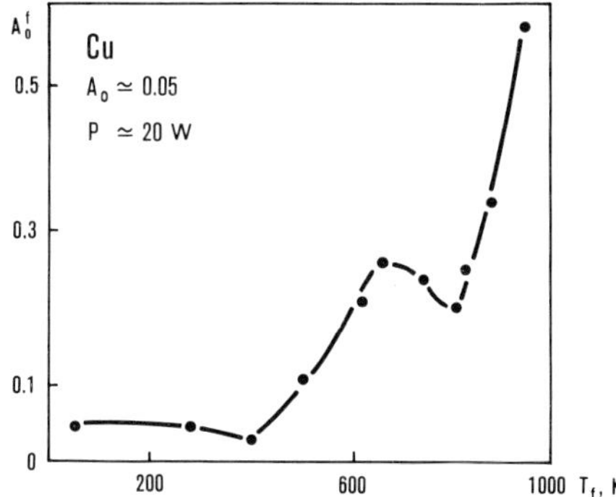

Figure 6.21 Cold absorptivity, A_0, of several copper foil samples, laser-heated in air up to various temperatures, T_f.

laser radiation (of low power, thereby avoiding superheating of the target), with a wavelength, λ_p, much lower than that of the main laser source. This method seems to be suitable for the study of oxide thickness during the heating stage prior to oxidation activation, when $x < 1\ \mu m$, that is $x < x_{max}^1$ (for $\lambda = 10.6\ \mu m$). As an example we show in figure 6.22 (i) the temperature–time variation of a titanium foil heated in air under the action of cw CO_2 laser radiation; (ii) the variation of sample reflectivity for a wavelength $\lambda_p = 0.6328\ \mu m$ (a He–Ne laser was used as a probe); and (iii) the theoretical values of the oxide layer thickness prior to the moment of oxidation activation [243].

So far we have examined only the simplified situation when a film consisting of a single oxide type is growing on the metal surface. For example, in the case of copper a monolayer system, $Cu + Cu_2O$, has been considered, while actually the oxide forming on copper may consist of two components,

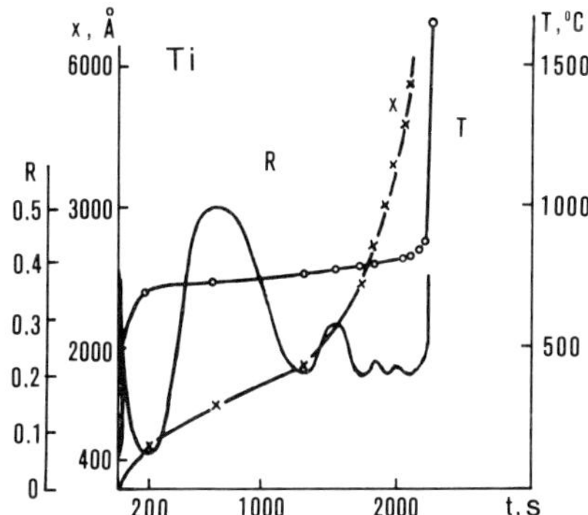

Figure 6.22 Kinetics of variation of the temperature, of the reflectivity at $\lambda \simeq 0.63\ \mu m$ and of the thickness of the oxide layer in the centre of a titanium foil sample during cw CO_2 laser irradiation.

$Cu_2O + CuO$ (the most oxygen-saturated oxide is situated in the proximity of the solid–gas interface). Even more complex is the structure of the irradiation area on iron samples, and on its alloys ($FeO/Fe_3O_4/Fe_2O_3$). On the other hand, for $T > T_a$, the oxide growing on zirconium and titanium samples consists, as a rule, of a single component (ZrO_2 and TiO_2, respectively).

The structure of a certain interaction area, which is strictly required in order to develop adequate theoretical interpretations of oxidising metal laser heating, can be supplied only by modern methods of surface analysis. Unfortunately, even for the activation stage of laser oxidation, these studies are only just beginning. A study of the structure and content of the oxide layer on laser-irradiated copper samples has been performed by replica extraction and subsequent electron diffraction investigation [59]. The experimental conditions are presented in table 6.2. Here T is the maximum temperature reached on samples by laser heating, and A_0, A_f, are the absorptivities of the samples before and after laser irradiation.

These results [59] allow a few conclusions. First, together with the growth in the maximum temperature ($T_f > T_a$), the oxide crystallite sizes also increase and, accordingly, the film thickness increases (a fact also confirmed by the evolution of the absorptivity A_f). Secondly, unlike the case of thin oxide film growth ($x \leqslant 1000$ Å), when the oxide film consists of Cu_2O only, in the oxidation activation stage we are faced with a two-component oxide layer.

It is obvious that the passage to a multilayer oxide makes the theoretical description of the process more difficult and, therefore, numerical solutions have so far been obtained only for the two-component system ($CuO + Cu_2O$) [224, 242]. In this case, according to the Wagner–Valensi theory [197–199, 201], the kinetic equation (6.2) takes the form

$$\frac{dx_1}{dt} = 2\frac{d_1(T)}{x_1} - \frac{M_1\rho_2}{M_2\rho_1}\frac{d_2(T)}{x_2}$$

$$\frac{dx_2}{dt} = 2\frac{d_2(T)}{x_2} - 2\frac{M_2\rho_1}{M_1\rho_2}\frac{d_1(T)}{x_1}. \qquad (6.30)$$

The indices 1 and 2 refer here to the oxides Cu_2O and CuO, respectively; ρ_1, ρ_2 are the densities of the two oxides, M_1 and M_2 their molecular masses,

Table 6.2 Data for some laser oxidised samples.

P (W)	t_f (s)	T_f (°C)	A_0 (%)	A_f (%)	Oxide	Crystallite size (μm)
27	79	370	1.1	2.5	$Cu_2O + CuO$	0.02–0.05
32	112	485	1.25	4.6	$Cu_2O + CuO$	0.05
31	105	505	0.75	9.9	$Cu_2O + CuO$	0.05–0.1
27.5	204	800	1.3	26	$Cu_2O + CuO$	0.1 –0.2

and d_1, d_2 the corresponding diffusion constants. Besides, the sample absorptivity depends on the values x_1 and x_2, but the actual expression (even approximated from [242]) of the dependence $A(x_1, x_2)$ is extremely intricate when interference phenomena are taken into consideration.

The numerical solution of the problem has shown that, for copper heating under the action of cw CO_2 laser radiation, the two-component structure of the oxide can be ignored and good accuracy can be obtained with the monolayer model ($Cu + Cu_2O$), which is far simpler. We have, however, to draw attention to the fact that for other metals or wavelengths situations may appear when consideration of the burnt multilayer structure becomes unavoidable.

Thus, for certain metallic foils heated under the action of a radiation generated by a YAG:Nd^{3+} laser source, the evolution of the interference diagram $A(x, t)$ is much more intricate than in the case of heating under the action of cw CO_2 laser radiation. In fact, it has been noticed that the high-frequency oscillations in the absorptivity $A(t)$ have been pre-modulated at a lower frequency. The diagram of the double interference observed finds an adequate interpretation only if one takes into account the two-component structure of the interaction area. In this case, the high-frequency oscillations in the absorptivity seem to be determined by Cu_2O oxide growth, while the low-frequency oscillations would correspond to the growth of the CuO oxide, whose optical constants at the wavelength $\lambda = 1.06\ \mu m$ are considerably different [244], while x_1, x_2 are of the order of the wavelength (see figure 6.23).

At the end of this section, let us mention that interference phenomena in the metal–oxide system not only permit an effective control over the growth of the oxide films (among others, in the case of non-laser heating) by recording the laser radiation absorptivity and/or reflectivity, but they also strongly influence the oxidation rate under laser heating. Thus, in the proximity of

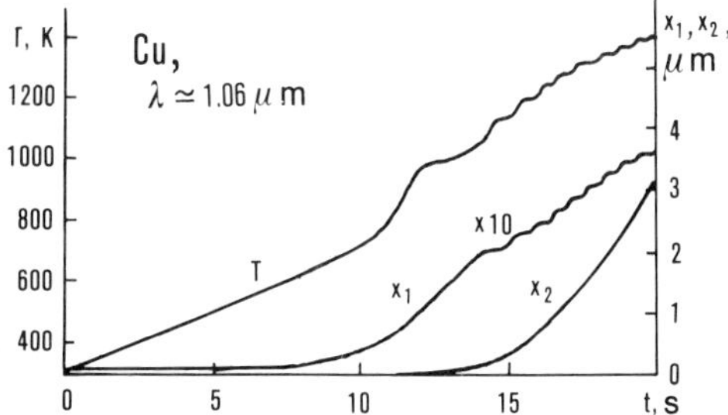

Figure 6.23 Time dependence of the temperature and thickness of the oxide layers when heating a copper sample under the action of a YAG:Nd laser. x_1 and x_2 are the thicknesses of the layers of Cu_2O and CuO, respectively.

the values A^k_{max}, the growth rate of the oxide film increases, and in the proximity of the values A^k_{min}, it decreases.

6.7. Ignition and burning of metals

Let us discuss now another important aspect concerning the effect of oxidising atmospheres upon the mechanism of metal heating under the action of laser radiation, namely heat release through the chemical reaction of oxidation—an aspect not considered in the above-mentioned experiments.

In the case when the thermally released power in the reaction, P_{ex}, obtainable by means of equation (6.10), is close to or exceeds the target power absorption, AP, the so-called metal burning regime occurs. As P_{ex} depends exponentially on target temperature (through the intermediary of dx/dt), the transition towards the burning regime occurs, as a rule, very quickly after a certain critical temperature, T_{ig} (known as ignition temperature), is reached.

Laser burning regimes of metals can be divided into two categories:

(i) Forced burning, when the energy conservation equation has the form

$$PA(T) + P_{ex}(T) = Q(T) \tag{6.31}$$

that is the burning (considering also the thermal power losses $Q(T)$) cannot be sustained except under the action of laser radiation.

(ii) Self-sustained burning, when the laser is merely playing the part of a 'trigger', ensuring target heating up to the ignition temperature, after which the P_{ex} value becomes so high that even without further laser energy dissipation it becomes possible to keep the target at temperatures for which the condition

$$P_{ex}(T) \geqslant Q(T) \tag{6.32}$$

is met.

Metal ignition under the action of laser radiation has been experimentally studied in [190, 195, 196, 204, 205, 214, 226, 231, 245–255].

A laser method for direct determination of P_{ex} has been proposed in [214], based on the values $A_e(T, t)$ and $R(T, t)$, and proceeding from the equation

$$\frac{P_{ex}}{P} = A_e - (1 - R). \tag{6.33}$$

For copper targets heated in air, it has been noticed that over the whole temperature range, $T_0 \leqslant T \leqslant T_m$, the condition $A_e \simeq A \simeq (1 - R)$ is rigorously met (see figure 6.18). Therefore copper heating, even over a relatively high temperature range, does not induce metal ignition.

A different situation occurs with laser heating of other metals such as titanium and zirconium [214, 251–255]. In figure 6.24 we have combined the characteristic curves $T(t)$, $A(t)$ and $1-R(t)$, corresponding to the laser–air-oxidation stage at high temperatures, in the case of a titanium target. As one can see, for $T \geqslant 1530$ K, the curves $A(t)$ and $1-R(t)$ diverge, a fact that indicates ignition of the metal target. Moreover, the difference $A_e - A$ reaches a value of $\simeq 0.4$ at a temperature $T \simeq 1750$ K, and its maximum value exceeds unity. In the same figure is displayed the time evolution of the power density thermally dissipated in the oxidation reaction, P_{ex}/S, that reaches a maximum of 20 W cm^{-2}. Forced burning takes no more than one second, a fact that can be accounted for in several ways. First, at sufficiently high temperatures and rates of the chemical reaction, the oxidation process begins to be limited by the rate of access of oxygen to the reaction area to a greater extent than by the diffusion of the reaction partners through the oxide layer. Accordingly, the value dx/dt is no longer obtainable from equation (6.2) but by means of the relation (6.3). Secondly, under the experimental conditions reported in [214, 251–255], the oxide layer is not peeled off the metal surface, and as its thickness x increases the rate dx/dt decreases at any temperature, T, that is reached by laser heating. The maximum temperature reached in the burning regime, in the absence of thermal conductivity losses, results from $P_{ex}(T_{max}) + PA(T_{max}) \simeq \sigma_{SB}\sigma_0(T_{max})T_{max}^4 S$.

After reaching T_{max} there is a certain drop in the target temperature, caused by a diminishing of P_{ex} with increasing x (as an effect of the diminishing dx/dt) as well as by enhancement of the blackening degree (emissivity), σ_0.

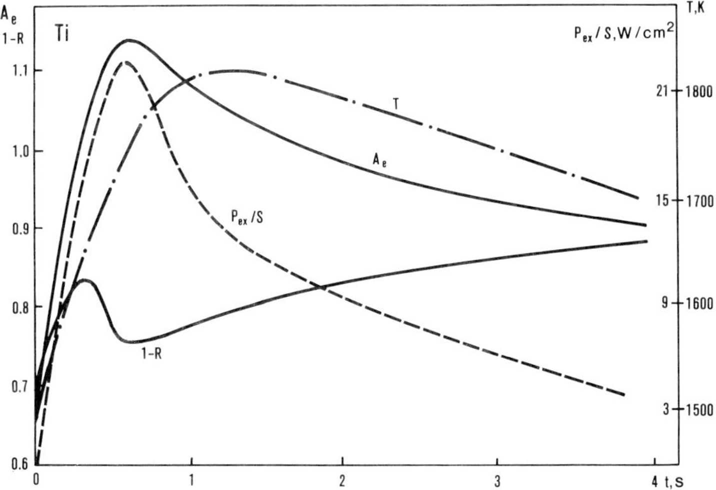

Figure 6.24 Kinetic curves of combustion in air for titanium foil under the action of cw CO_2 laser radiation.

In order to initiate self-sustained burning (and, in general, to increase the P_{ex} value), it is necessary, on the one hand, to enhance the access of oxygen to the irradiated area and, on the other hand, to provide for removal of the reaction products. The first of these requirements can be met by blowing an air/oxygen jet upon the irradiation area (in the same way as is usually done in the case of metal gas jet cutting). The proper fulfilment of the second requirement depends on the metal. From this viewpoint, metals can be divided into the following groups:

(i) Case $T_{ig} < T_m$. There are two possibilities: (*a*) The target burning temperature, T_b, is higher than the temperature initiating intense oxide vaporisation T_v^{ox}, and, as a result, oxide thickness remains insignificant, allowing for P_{ex} to be sustained at high levels. This regime is characteristic of metals with a high melting point, such as tungsten and molybdenum, and can be used for laser-induced chemical etching [196]. (*b*) T_{ig}, $T_b < T_v^{ox}$, but the oxides are in poor mechanical contact with the metal base (spongiform interaction area), and can be removed from the target surface by means of a gas jet. Such a situation occurs for laser burning of titanium at moderate levels of incident laser intensity, I_0.

(ii) Case T_{ig}, $T_b > T_m$. The gas jet causes the liquid metal to be removed from the reaction area, together with the formation of oxide on its surface. Copper, aluminium and their alloys as well as different types of steel belong to this group.

Let us illustrate these by some experimental examples.

The self-sustained burning regime in an air jet has been obtained through multipulse laser irradiation of a titanium foil [231]. The average radiation power was $P \simeq 60$ W, the focal spot radius $R_s \simeq 3.3$ mm and the metal foil thickness $h \simeq 50$ μm. Foil melting did not occur. The air jet was directed along the target surface at speeds up to 70 m s^{-1}. It was noticed that for a speed $v_{opt} \simeq 11$–14 m s^{-1}, after the target reached the temperature T_{ig}, the laser could be turned off while the foil continued to burn in the air jet. Moreover, the burning front got out of the irradiation spot and moved along the direction of the air jet at a speed of 0.1–1 cm s^{-1}. For all the other values of v only forced burning occurred, punching a hole by the removal of the reaction products.

The existence of an optimal value of the gas-blowing speed is determined by competition between the above-mentioned positive (from the burning point of view) functions of the gas jet and its cooling action upon the target. This competition has been clearly emphasised in the literature concerned with gas jet laser-cutting techniques. In figure 6.25 the evolution of the specific energy consumption per unit length of cutting, P_1 (W mm^{-1}), as a function of the speed of laser-cutting of stainless steel, v, is plotted for two types of gas (inert and oxygen). Figure 6.26 shows the dependence of the speed v on the pressure p of the oxygen from the gas jet directed perpendicularly onto the

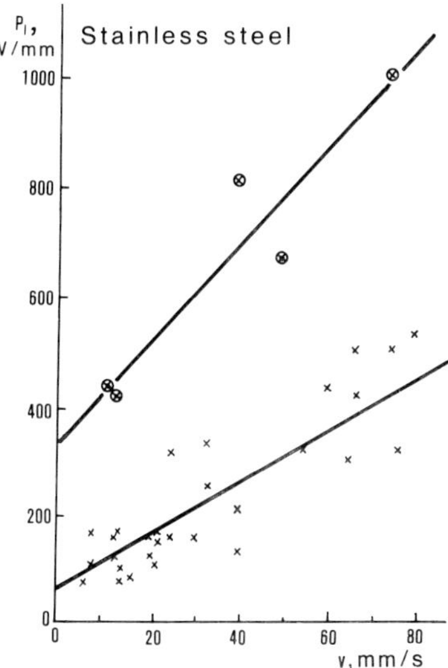

Figure 6.25 Evolution of specific energy consumption per unit length of the cut, P_l, as a function of the speed of the stainless steel laser cutting sustained by an inert gas jet ($\otimes$), and by an oxygen jet ($\times$), respectively.

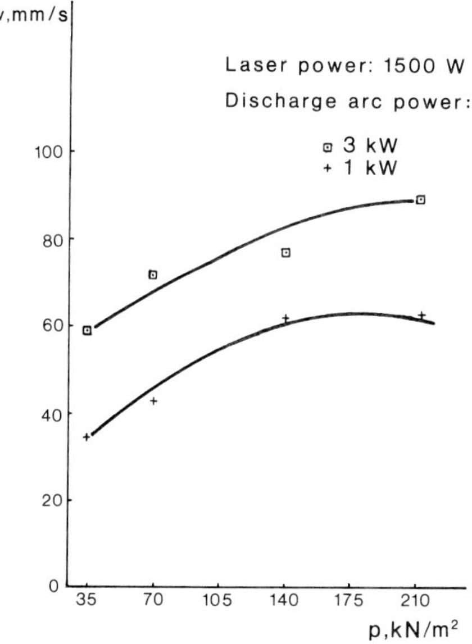

Figure 6.26 Effect of oxygen pressure on cutting speed in arc-augmented laser cutting.

irradiated surface [83]. It can be seen that the use of a chemically active atmosphere results in reduced energy demands for metal thermal processing and that there is an optimum pressure, p_{opt}, or as it has also been said, an optimum value of the gas-blowing speed, v_{opt}, at which the forced burning regime becomes more efficient.

Metal burning (titanium, various steel compositions), at $T_b > T_m$, with a gas blown along the surface, has been studied in reference [247]. The static pressure into the jet was $\simeq 1$ atm, and gas speed, $v \simeq 0.8$ M. The targets had a cylindrical shape and were rather long. Irradiation took place at the end of one of the targets (cw CO_2 laser radiation, at $\bar{I} \simeq 2$–$16\ \mathrm{kW\,cm^{-2}}$). Such an experimental geometry has eliminated to a considerable extent all kinds of losses in the irradiation spot, except for thermal conductivity along the metallic cylinder, and has allowed for the evaluation of the ablation speed, v_a, of melt and oxides, during the irradiation process. After ignition, the burning process evolved at a constant target surface temperature, $T_b \simeq$ const, as measured pyrometrically. Some of the results of the investigations performed in an air jet are given in table 6.3.

We note a rather unexpected outcome resulting from the study of the data in table 6.3. Thus, together with the variation in the incident laser intensity, $\bar{I}$, only the metal and oxide ablation speed are changing. Values of the burning stationary temperature and of the effective power density thermally dissipated in the oxidation reaction $(P_{ex} - Q)/S_s$ (i.e. after extraction of the power of thermal losses) do not depend on $\bar{I}$.

We mention that the last of these quantities is comparable with the value of the laser radiation intensity dissipated into the metal, and therefore represents an efficient additional heating source.

Metal burning in the oxide vaporising regime has been studied in reference [226]. Thermally thin and thermally insulated tungsten foils were irradiated in air, at intensity levels of $\bar{I} \sim 10^2\ \mathrm{W\,cm^{-2}}$. Typical CRO traces of the signals $T(t)$, $\mathrm{d}T(t)/\mathrm{d}t$ are reproduced in figure 6.27. The moment of ignition and the temperature related to it, $T_{ig} \approx 600$–$800\ °\mathrm{C}$, have been determined on the basis of the vapours suddenly brightening, which occurs when reaching a maximum in the $\mathrm{d}T/\mathrm{d}t$ derivative. Complementary experiments have indicated that the evaporated substance is WO_3. Tungsten burning is of the

Table 6.3 Data on the laser burning of some metals in air.

Metal	$v_a = f(I)$ (mm s^{-1})	T_b (K)	A (%)	$(P_{ex} - Q)/S_s$ (kW cm^{-2})
Stainless steel 304	$1.3 + 0.58\,\bar{I}$ kW cm^{-2}	2450 ± 100	78 ± 6	1.7 ± 0.9
Stainless steel 310	$0.8 + 0.46\,\bar{I}$	2400 ± 100	51 ± 5	0.9 ± 0.7
Stainless steel 1020	$0.4 + 0.45\,\bar{I}$	1980 ± 50	50 ± 4	0.4 ± 0.5
Titanium	$1 + 0.46\,\bar{I}$	2300 ± 70	38 ± 5	0.8 ± 0.5

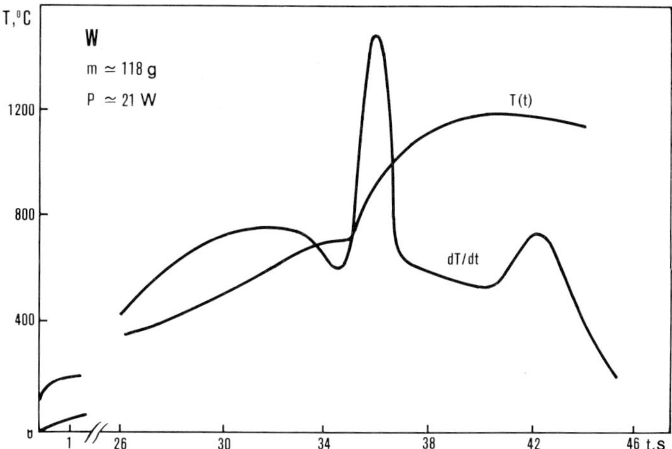

Figure 6.27 Experimental oscillograms of the temperature signal, and of its derivative, $\mathrm{d}T/\mathrm{d}t$, corresponding to the laser-induced combustion of a tungsten target.

forced kind. On the $\mathrm{d}T/\mathrm{d}t$ curve certain interferential oscillations can be noted, revealing the dynamics of the variation in thickness of the oxide layer. From the first peak it is possible to state with precision that at the moment of ignition one had $x \approx \lambda/4n \approx 1\ \mu\mathrm{m}$. The oxide keeps growing until an upper limit thickness is reached. Then, after the stationary burning regime takes control ($T_b \simeq$ constant), the reverse can be observed, that is the thickness decreases as an effect of the vaporisation of the oxide layer. The existence of a third peak in the $\mathrm{d}T(t)/\mathrm{d}t$ evolution is proof that the oxide layer thickness decreases again as a result of ablation, down to $x \leqslant 1\ \mu\mathrm{m}$.

Therefore, for laser burning of tungsten (and this is also true for molybdenum), a continuous process is taking place, in which the front of an oxidation wave propagates deep into the sample, its speed being determined by the actual energy dissipation into the sample (that is, taking into account thermal power losses). Behind the wavefront moving along the normal to the target surface (yet at a certain distance, $x \leqslant 1\ \mu\mathrm{m}$), the substance appears in the form of a vapour–air mixture.

This pattern suggests that a relatively simple analytical approach [226] can be used to understand the peculiarities of laser ignition and burning of metals. At sufficiently high temperatures, the kinetic equation takes the form

$$\frac{\mathrm{d}x}{\mathrm{d}t} = \frac{d_0}{x}\exp\left(-\frac{T_\mathrm{d}}{T}\right) - v_0\exp\left(-\frac{T_\mathrm{v}}{T}\right) \tag{6.34}$$

where v_0 and T_v are the constants from the oxide vaporising law. Similarly, the energy conservation equation for a thermally thin foil (6.9) is transposed to the case of laser burning in the form

$$cm\frac{\mathrm{d}T}{\mathrm{d}t}=PA(T)-Q(T)+\rho SW_{\mathrm{H}}\frac{d_0}{x}\exp\left(-\frac{T_{\mathrm{d}}}{T}\right)-\rho Sqv_0\exp\left(-\frac{T_{\mathrm{v}}}{T}\right) \tag{6.35}$$

where q is the latent heat of oxide vaporisation.

For the stationary burning regime, $\mathrm{d}x/\mathrm{d}t=0$, one has $x=(d_0/x_0)\times\exp[(T_{\mathrm{v}}-T_{\mathrm{d}})/T]$. By replacing this value of x in (6.35), and for $\mathrm{d}T/\mathrm{d}t=0$, one obtains the following relation, which now permits determination of the stationary temperature

$$PA(T)+\rho Sv_0(W_{\mathrm{H}}-q)\exp\left(-\frac{T_{\mathrm{v}}}{T}\right)=Q(T). \tag{6.36}$$

For instance, in the case of WO_3, the difference $W-q$ is positive and we have a ratio between powers of chemical reaction heat release and thermal losses typical for laser ignition of metals (figure 6.28). In the diagrams the instability point $T=T_{\mathrm{ig}}$ corresponds to the ignition temperature. After a further small rise in temperature self-acceleration of the reaction begins, finally followed by a stationary regime ($T=T_{\mathrm{b}}$) which corresponds either to self-sustained burning ($P=0$), or to forced burning. Curves 1 and 2 would correspond to self-sustained burning and 1 and 3 to the case of high thermal losses, when burning is no longer possible at $P=0$. Curves 5 and 7 and 6 and 7 demonstrate the feasibility of obtaining forced burning, and curves 4

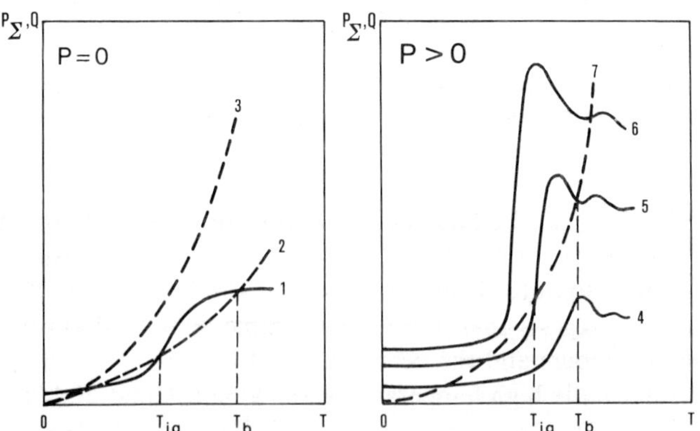

Figure 6.28 Dependence on temperature of the power, P_Σ (full curves 1, 4–6) and of the thermal power losses, Q (broken curves 2, 3, 7), in the case of a laser-oxidised metal for various values of the thermophysical constants of the process. ($P_\Sigma=AP(T)+P_{\mathrm{ex}}(T)$).

and 7 correspond to the case when the absorbed laser power and the heat released in the chemical reaction are not sufficient to compensate for the thermal losses within the target temperature range where the activation of the oxidation reaction is taking place.

Let us emphasise that if the incident laser power is sufficient to sustain forced burning, but not high enough to heat the metal up to the ignition temperature, an interesting situation occurs, as reported in references [241, 242, 256, 257]. Metal ignition can be initiated in this case through the pulsed action of an additional laser.

A number of peculiarities of metal burning under the action of laser radiation have been examined in [213, 248–250, 258, 259]. Thus [250], using quantitative methods based on non-linear collision theory, it has been demonstrated that when irradiating metals with high melting points there may occur not only a stationary laser burning regime, but also a self-oscillation burning regime. Different kinds of 'echoes' of the burning process on external perturbations (stand-by and control trigger regimes) have been analysed. In a theoretical paper [258], the existence of certain critical values for the radiation parameters (I_0 and R_s) has been established. When these critical values are topped, the temperature field becomes unstable and the burning front may propagate outside the irradiated area on the target surface.

Experimental evidence of 'stochastic laser burning', initiated by the strong focusing of cw CO_2 laser radiation on titanium foils, is reported in [248]. The initiation of complex stochastic oscillations in the final burning stage, that is when the target temperature decreases, has been emphasised, while these oscillations were absent in the case of a sample uniformly illuminated. The authors ascribed their experimental results to the excitation of a special regime (chaotic evolution dominated by a 'strange attractor') as a result of the initiation of radial heat elimination. Nevertheless, it was shown [214, 259], that under the experimental conditions reported in reference [248], the temperature reached in the irradiated area was different from that on the opposite side (the evolution of which was studied by means of thermocouples) and surpassed the melting point of the metal, T_m. A completely different mechanism has been suggested for this finding [213, 259], providing an explanation for the oscillations in $dT(t)/dt$ observed in [248]. It is based on the phase transformations of titanium, i.e. melting and crystallisation, and is not related to the oxidation process. In the theoretical study [260], the problem of the temperature oscillations of the type reported in [248] is considered strictly from the thermochemical point of view. It is shown that this instability can be related to a delay in the chemical reaction's thermal effect upon the temperature change. In fact, for a quick rise in temperature, the surface concentration of metal ions diminishes and their diffusion flux towards the oxide surface increases. At the initial moment this flux is maximal, then it decreases down to a value that depends on the electron concentration profile corresponding to the enhanced temperature. The value P_{ex} also varies in the same way. Accordingly, initially a situation could be created in which P_{ex} exceeds the power of the thermal losses, thus determining a further rise in temperature. The duration of such oscillations is of the order of the diffusion time characterising the establishment of the concentration profile in the oxide layer.

Chapter 7 Optical Effects and Diagnosis of Thermochemical Interaction Processes

The discussion of the thermochemical processes continues, dwelling upon specific optical effects that develop in characteristic non-equilibrium systems. On these grounds new applications are introduced that use dynamic thermochemical laser methods to determine various constants of metal oxides, and possibly also of other compounds.

One peculiarity of laser oxidation and combustion of metals consists of the pronounced variation in the optical properties of the metal–oxide system during irradiation. In this chapter we shall discuss these effects and show that the growth in target absorptivity (often non-monotonic) after a chemical reaction has been activated not only determines the efficiency of metal processing by laser beams in an oxidising atmosphere, but is also a phenomenon that can be used for a diagnosis of the surface chemical reaction kinetics, and for determining the optical constants of the reaction products. This means that the laser is not only a clean, powerful and localised source of thermal energy that can be delivered to the investigated/processed metal sample, but is also an interesting new tool to induce high-temperature surface chemical reactions.

7.1. Optical effects in non-equilibrium systems

By now we have examined several mechanisms determining the deviation of $A(T)$ from A_0. Let us summarise:

(i) The effect of laser surface cleaning [211] occurring at $T < T_a$ and $A_0 > A_M$.

(ii) At moderate heating rates (again at $T < T_a$), the change in the absorptivity $A(T)$, can be the effect of the variation with temperature of the metal absorptivity $A_M(T)$ [23] (that has to be taken into consideration in all the stages of the heating process), of the dissolution of oxygen into the metallic foil [230] and/or of changes in the stoichiometric content in the case of alloys [235].

(iii) The absorption of laser radiation into the oxide layer (in particular the interferential oscillations) [208] of $A(T)$, at $T > T_a$.

To these one may add:

(iv) The influence on the absorption of radiation into the oxide layer of the increase in concentration (with the increasing temperature) of mobile charge carriers in the oxide [208, 227, 261].

(v) The variation of the optical properties of the target during intense metal combustion when the oxidation reaction becomes gas diffusion limited and such phenomena as the modification of the oxide stoichiometric content [251, 252], oxide film dissolution [253] and formation of other compounds, different from the oxides (such as nitrides or oxynitrides [254, 255]), can take place.

Let us now discuss more thoroughly these latter mechanisms ((iv) and (v)), that may generate a non-monotonic variation in the absorptivity $A(T)$, at $T > T_a$.

The conjecture has been made [203], that in the case of metal oxidation under the action of cw CO_2 laser radiation, the non-monotonic behaviour of the absorptivity $A(T)$ for a large rise in temperature, could be not only the effect of the interference phenomena in the metal–oxide system, but also a result of the plasma resonant action upon the free carriers in the oxide layer.

Indeed, at room temperature, the plasma frequency

$$\omega_p = ((4\pi n_e e^2)/m_e^*)^{1/2} \tag{7.1}$$

(where e, m_e^* are the electron charge and electron effective mass, respectively, and n_e is the electron concentration) in the oxides is usually low, as compared with the frequency of the CO_2 laser radiation, ω. Yet, during the laser heating process the value of the electron concentration in the oxide layer increases and, when reaching a certain temperature, the condition $\omega \simeq \omega_p$ can be met. As a consequence, the reflection of laser light on the oxide film is strongly enhanced, the absorption decreases accordingly, and the temperature decreases too.

Experimentally, such an anomalous behaviour of $A(T)$ has been observed in [227] by heating in air several vanadium targets under the action of cw CO_2 laser radiation. The evolution of $A(T)$, experimentally recorded in the stage of high-temperature heating and oxidation, can be followed in figure 7.1 (curve 1). The minimal absorptivity, A_{min}, at $T \simeq 1200\ ^\circ C$, corresponds to resonance with the plasma frequency in the oxide layer.

The possibility of clearly observing this effect is related to the anomalous properties of V_2O_5, which is the main constituent of the oxide layer after the activation of the chemical reaction in air [229].

First of all, at high temperatures, the oxide is forming in the liquid phase, as its melting point is at only $\simeq 680\ ^\circ C$. Also, it has a high absorption index $k \sim 1$ (corresponding to the laser radiation wavelength $\lambda = 10.6\ \mu m$) and a significant thickness ($x \sim 100\ \mu m$). As a result, the radiation does not in fact reach the metallic foil (being completely absorbed in the oxide layer) and the interference phenomena in the metal–oxide system are insignificant. The value of the absorptivity, $A(T)$, is thus determined in the given temperature range by the optical characteristics of the oxide only, i.e.

$$A = \frac{4n}{(n+1)^2 + k^2}. \qquad (7.2)$$

Secondly, at room temperature near $\lambda = 12\ \mu m$ there is an absorption band for V_2O_5 (in figure 7.2 the functions $k(\lambda)$, $n(\lambda)$ and $A(\lambda)$ are plotted).

In the literature data concerning the optical properties of V_2O_5 in the temperature range of interest, $T \geqslant 1100\ ^\circ C$, are scarce. Yet, it seems reasonable to assume that, as temperature rises, the radiation wavelength for which a minimum absorptivity is recorded, λ_{min}, would decrease, so that the resonance with the plasma frequency could be noticed at $\lambda \simeq 10.6\ \mu m$. Indeed, the number of free carriers and the oxide conductivity rise exponentially with

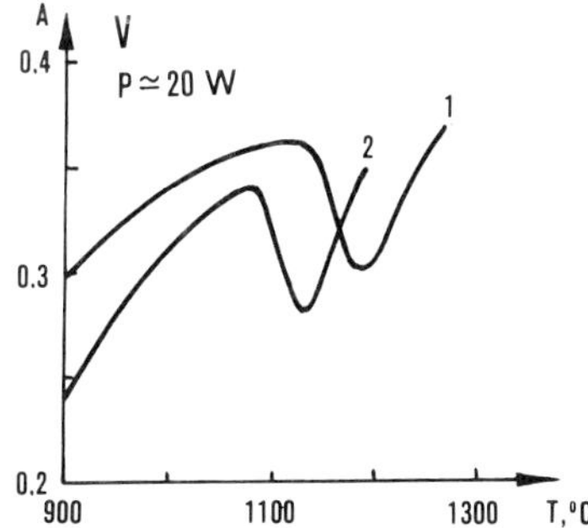

Figure 7.1 The dependence on temperature of absorptivity of a vanadium sample, $A(T)$, during cw CO_2 laser heating $P \simeq 20$ W, without additional illumination (1), and with illumination by means of a mercury-vapour lamp (2).

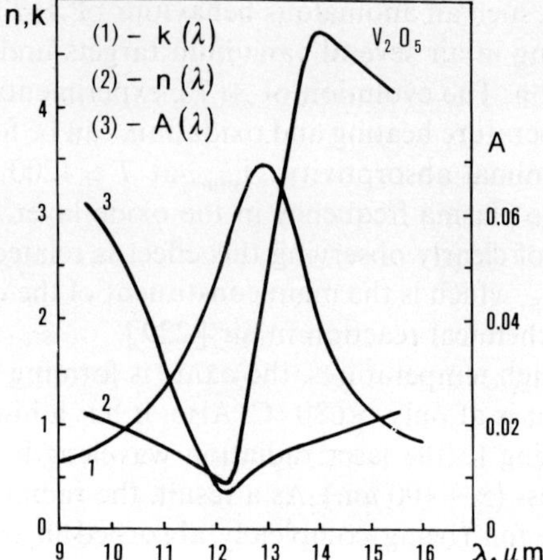

Figure 7.2 Spectral dependences of absorptivity $A(\lambda)$, absorption index, $k(\lambda)$, and refraction index, $n(\lambda)$, for V_2O_5 at room temperature.

the temperature (according to data in reference [262] the increase in the conductivity of V_2O_5 reaches four orders of magnitude).† In order to test the hypothesis according to which of the peculiarities observed in the behaviour of the absorptivity of laser-oxidised vanadium, in air at $T \sim 1100$–1200 °C, are related to an increase in the electron concentration, control experiments were performed [227]. The samples were illuminated by means of a mercury lamp, the main fraction of the emitted radiation thus being in the UV spectral range. As the forbidden gap of V_2O_5 is narrower than the energy of the UV quanta, a high increase in the electron concentration of the conduction band is expected to be induced through the photoelectric effect, as an effect of the irradiation. This has been confirmed by direct measurements of the conductivity of the oxide layer. As a result, a shift of $A_{\min}$ towards lower temperatures (see figure 7.1, curve 2) has been noticed. The corresponding value of the temperature shift for the minimum absorptivity as found in different experiments is situated in the range 20–100 °C.

Let us note that the effect, reported in reference [227], of resonance with the plasma frequency upon laser-oxidised target absorptivity, represents a singular case at $\lambda = 10.6\ \mu$m, although by the use of tunable lasers analogous effects can also be noticed with other metallic oxides. However, in the case of CO_2 lasers, the temperature-dependent increase in the electron concentration generally only brings some corrections to the value of $A(T)$, as an effect of the corresponding changes in the oxide's optical constants.

† We mention that near the plasma frequency an important thermal electromotive force between the metal and the oxide can be generated as an effect of the thermal diffusion of charge carriers [263].

An interesting anomaly in the absorptivity behaviour has been noticed in the stage of laser burning of titanium targets in air [211, 231, 251]. Thus, on the $A(t)$ curve plotted in figure 6.24, one can see that through target heating in the $T \simeq 1600-1800$ K temperature range, a sharp oscillation is recorded. In other words, initial reaching of a maximum value $A_{max} \simeq 0.85$ is immediately followed by a minimum value, $A_{min} \simeq 0.74$.

In reference [231] this oscillation has mistakenly been ascribed to interference in the metal–oxide system. By the processing of A_{max} and A_{min} values assuming that these values correspond to the first interferential maximum and, respectively (see below), minimum in the absorptivity evolution, it has resulted that $n = 1.41$ and $k = 0.07$, in good agreement with the known values of the two coefficients for TiO_2 [264]. As we shall show, the oscillatory nature of the absorptivity behaviour of the titanium targets during forced burning and in conditions of the temperature dependence of the oxide layer absorptivity, $A(T)$, cannot be explained on the basis of interference in the metal–oxide system (although certain numerical calculations [261] seem to hint at such a possibility).

As has been shown [251, 252], the key to the understanding of the laws governing the variation of the optical characteristics of the targets during their laser burning, when reaction rates are high, is to consider the limitation of the reaction rate as an effect of mass transport in the gaseous phase. In this way a direct connection between the optical properties of the oxidising metal and the chemical kinetics can be established.

The mechanism proposed in reference [251] with a view to interpreting the anomalous behaviour of $A(T)$ for titanium is based on the fact that the oxide optical constants, n and k, do not depend only on the oxide conductivity, $\sigma(T)$, but also on its deviation z from stoichiometry (in the case of titanium oxidation, TiO_{2-z}). Indeed, the non-stoichiometric oxides (having oxygen vacancies) occur precisely in the case of high reaction rates specific to the metal laser burning regime.

As is known [200], with non-stoichiometric oxides, the variations of z and σ are interdependent, and normally $\sigma \sim z$. On the other hand, inside the interaction area there is always a gradient in the charge carrier concentration, $\mathrm{d}n_e/\mathrm{d}\xi \neq 0$ (ξ is the coordinate along the direction perpendicular to the metal surface). Accordingly, there will also be a conductivity gradient, $\mathrm{d}\sigma/\mathrm{d}\xi \neq 0$, and consequently a spatial inhomogeneity of the optical properties of the oxide (on the direction of the vector $\boldsymbol{\xi}$) is to be expected. At moderate heating rates, this inhomogeneity can be neglected (as we have done when calculating the $A(T)$ behaviour in the case of some copper targets for $T < T_m$). Yet in the laser burning process the gradients $\mathrm{d}n/\mathrm{d}\xi$ and $\mathrm{d}k/\mathrm{d}\xi$ become significant. Let us illustrate this with some numerical results.

Within the temperature range of interest, the degree of non-stoichiometry z of the TiO_2 oxide is determined by the Ti^{4+} interconnected ions. The carrier concentration will accordingly be $n_e = 2\,Nz$ (where N is the volume concentration of Ti^{4+} ions). At the oxide–air boundary ($\xi=\xi_1$), the value $z=z_1$ is maximum, and depends on the local temperature T_1 and on the oxygen partial pressure at the surface, p_w, $z_1=z_{10}p_w^{-1/5}\exp(-T_{10}/T_1)$. As is known, in the burning stage, with limitations on access of the gaseous phase, we have $p_w \ll p_{O_2}^{\infty}$, where $p_{O_2}^{\infty}$ is the oxygen pressure far off the target. At the other boundary, the value $z=z_2$ is determined by the equilibrium between TiO_2 and the intermediary oxides Ti_nO_{2n-1} [200], $z_2=z_{20}\exp(-T_{20}/T_2)$. The temperature inhomogeneity inside the oxide layer can be neglected ($T_1 \simeq T_2 \simeq T_w$). Then, in the quasistationary approximation, the transverse distribution of z in the oxide layer is linear

$$z(\xi)=z_2+\frac{(\xi-\xi_2)(z_1-z_2)}{x}. \tag{7.3}$$

Consequently, the conductivity profile $\sigma(\xi)=n_e e\mu=2Nz(\xi)e\mu$ will also be linear (μ means here the mobility of the carriers). According to data in reference [200], we have $T_{10}=24\,800$ K; $T_{20}=6900$ K, $z_{10}=475$ atm$^{1/5}$, $z_{20}=3.1$, $\mu=0.1$ cm^2 W^{-1} s^{-1}.

Essential for the model is the relation between the conductivity $\sigma(\xi_1)$ and the pressure p_w, wherefrom the direct link between the absorptivity of the oxidising target and the reaction rate follows. The pressure p_w can be obtained from the condition of equality between the Ti^{4+} cation flux through the oxide and the flux of O_2 molecules from the ambient air necessary for the oxidation process

$$Nd_0\exp\left(-\frac{T_d}{T}\right)\{1-z_{10}p_w^{-1/5}\exp[(T_{20}-T_{10})/T_w]/z_{20}\}/x=D_{O_2}\left(\frac{dn_{O_2}}{d\xi}\right)_w. \tag{7.4}$$

Here $d_0=330$ cm^2 s^{-1}, $T_d=33\,000$ K [265] are Wagner constants, $D_{O_2}=0.18$ $(T/273)^{1.724}$ cm^2 s^{-1} [92] is the coefficient of oxygen diffusion in air and $(dn_{O_2}/d\xi)_w=(p_0/T_0-p_w/T_w)/k_B\Lambda$ is the gradient of the concentration of O_2 molecules to the surface proximity (p_0 and T_0 are the oxygen partial pressure and the temperature of the gas far off the target, k_B is Boltzmann's constant and Λ the diffusion length (of the order of the irradiation spot size)). To the relation (7.4) one must add the energy balance equation (or that of the thermal conductivity in the case of the existence of temperature gradients inside the irradiation area) and the equation of the growth rate and the input impedance Z of the oxide layer

$$\frac{dx}{dt}=\frac{d_0}{x}\exp\left(-\frac{T_d}{T}\right)\{1-z_{10}p_w^{-1/5}\exp[(T_{20}-T_{10})/T_w]\}/z_{20} \tag{7.5}$$

$$\frac{dZ_{ex}}{dt}=-2\pi i(1-\varepsilon_c(\xi)Z_{ex}^2)/\lambda. \tag{7.6}$$

The local value of the absorptivity is related to Z_{ex} through the formula

$$A = 1 - [(Z_{ex}(\xi_1) - 1)/(Z_{ex}(\xi_1) + 1)]^2 \tag{7.7}$$

and the boundary condition has the form $Z_{ex}(\xi_2) = \varepsilon_M^{-1/2}$. Here ε_M is the dielectric permittivity of the metal base, which can be expressed by means of its absorptivity (i.e. of the saturation value $A_s > A_0$ in the case of titanium oxidation).

The calculations based on this model [251, 252] have indicated that for the most intense laser oxidation of titanium in air, when the intensity of the exothermal effect reaches 20–50 W cm^{-2}, access of the gaseous phase into the reaction is limited, and the oxygen partial pressure near the surface, p_w, suddenly decreases. As a result, the oxide conductivity tends to its maximum and k reaches a value slightly exceeding unity. These behaviours can supply a good explanation for the minimum in the absorptivity evolution, A_{min} (figure 3.64), that corresponds to the peak in the exothermal effect (at the highest rate of the chemical reaction) and not to the peak in temperature. There is also good agreement between the calculated and experimental curves describing the high-temperature heating in air of titanium. The corresponding curves reported in reference [251] are presented in figure 7.3 (instant 0 indicates the beginning of the high-temperature stage).

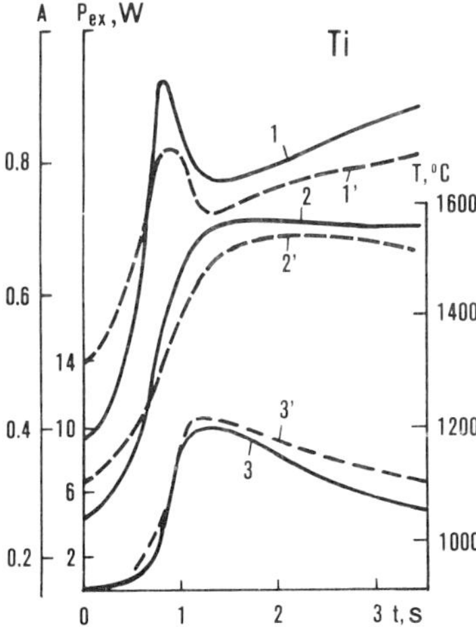

Figure 7.3 Experimental (broken curves) and calculated (full curves) time dependences of the absorptivity $A(t)$, (1, 1′) of the temperature in the centre of the sample's opposite face (non-exposed to laser irradiation), $T(t)$, (2, 2′) and of the power, P_{ex}, of the exothermal effect (3, 3′) (titanium cylindrical sample with a radius of 2 mm and a thickness of $\simeq 0.3$ mm).

Additional confirmation of the importance of the examined mechanism is offered by the experiments reported in reference [252], concerning the blowing of a weak jet of air onto the laser-irradiated target. Such a jet should not change the sample temperature and should not influence the evolution of $A(T)$ from a thermophysical point of view. However, a significant change in the A_{max} and A_{min} values has been recorded (for example A_{min} decreases down to 0.6 as compared to 0.75–0.8 previously) and both extremes move to higher temperature values (figure 7.4).

These effects cannot be interpreted within the framework of the optically homogeneous burning area model, with constant values for n and k, but can be explained by the limitation of access of the gaseous phase to the reaction. Indeed, when introducing the gas jet, the limitation regime is shifting towards higher temperatures that correspond to higher values of the oxide conductivity, σ, and to lower values of the absorptivity, A, respectively.

It is necessary to point out here that so far there is no direct experimental evidence of the mechanism discussed above, though in reference [267] it was shown that during burning of titanium in air the conversion of TiO_2 to TiO—a lean oxide as far as oxygen content is concerned—takes place.

In this way, based on the example of titanium ignition, it has been shown that through laser irradiation in a well determined regime, conditions can be created for a direct influence of the reaction rate and of the gas transport processes upon the optical characteristics of the interaction area and particularly on target absorptivity. One can naturally expect that this mechanism could be initiated for other metals also, the oxides of which can present significant departures from stoichiometry.

From this viewpoint, the metal closest to titanium is zirconium. Also in this case, an anomalous (and even more complex) behaviour of the absorptivity, $A(t, T)$, of some zirconium samples laser-irradiated in air has been recorded [254, 255], being apparently caused in the first place by the limitation of gaseous phase access. However, in this case the absorptivity behaviour is determined not only by a modified oxide stoichiometry, but also by the

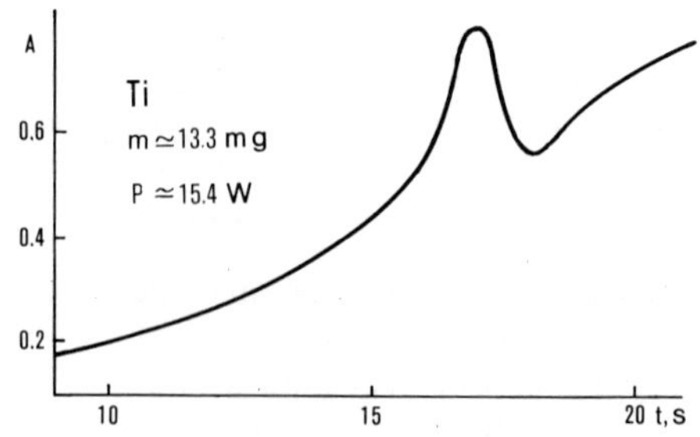

Figure 7.4 Absorptivity of a titanium sample ($m \simeq 13.3$ mg, $P \simeq 15.4$ W) during the continuous blowing of air over the laser irradiation region.

synthesis of another type of superficial compound, such as metal nitrides and/or oxynitrides.

Thus, for the first time [254, 255] the other chemically active constituent of air—nitrogen—was involved in the interaction of laser radiation with metals. Indeed, while in most cases the oxidation reaction prevails in the heating of metals in air, this does not preclude situations when the main part is played by the synthesis of nitride compounds. This is precisely the kind of situation that may occur in the high-temperature stage of zirconium burning, when near the reaction surface the oxygen concentration sharply decreases with the progress of metal oxidation, while the nitrogen concentration either remains constant or increases. Therefore, even at low reaction rates, determined by the values of d_0 and T_d, the nitridation reaction can prevail as compared with the oxidation one.

CRO traces recorded during high-temperature heating of some zirconium foils of successively lower masses are given in figure 7.5 (steady laser power, no air blowing). Generally, the dynamics of zirconium absorptivity is much

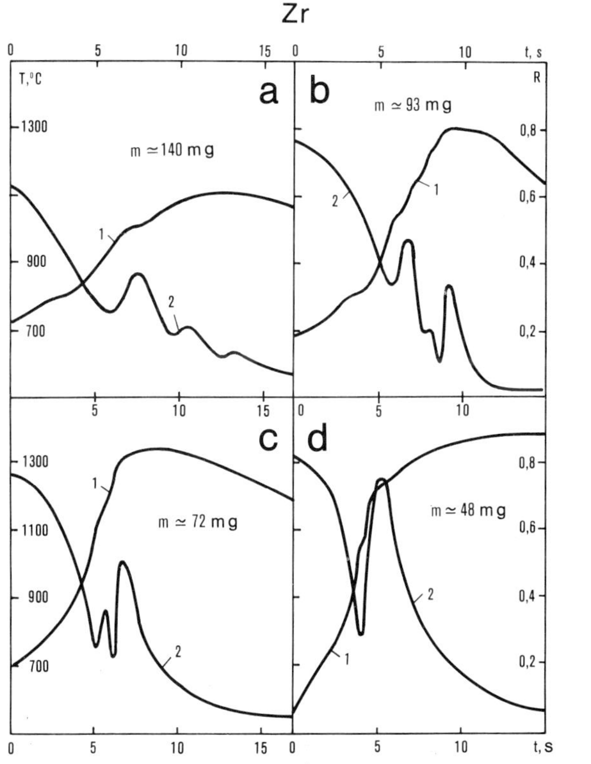

Figure 7.5 CRO traces of zirconium target heating: (1) $T(t)$, (2) $R(t)$. The sample masses were (*a*) 140 mg, (*b*) 93 mg, (*c*) 72 mg, (*d*) 48 mg.

more specific than that of titanium targets. It can be seen that the curves $A(t)$ can consist, in relation to the sample's mass, of two distinct regions: (i) an interferential one, where the thickness of the superficial layer is $x \geqslant \lambda/4n$ (in contrast with the case of titanium laser heating) and is still transparent enough for the radiation; and (ii) a second region corresponding to the situation when the laser radiation does not in fact reach the metal surface, but nevertheless, as a result of the access limitation of the gaseous phase, the absorptivity value, A, can go down to 0.15–0.2. While, at large enough values of the sample mass, and for a given value of the incident laser power, P, sample ignition may not occur, and in $A(t)$ only the first (interferential) phase of the process can be observed (figure 7.5(a)). On the other hand, at high heating rates, complete disappearance of the interferential stage in $A(t)$ is possible (figure 7.5(d)).

We mention that the extremely low values, $A \simeq 0.15$–0.2, recorded in zirconium laser burning at high temperatures, cannot be explained by resort to the dependence of oxide conductivity on temperature, $\sigma(T)$ and to its non-stoichiometry, as is the case with titanium targets. Moreover, these extremely low absorptivity values are characteristic of metals rather than semiconductors. Starting from the well known fact of the abnormally high conductivity of ZrN (as well as of the other mononitrides of the metals from the IVA subgroup) at room temperature, the authors of papers [254, 255] have tried to determine if a nitride layer is formed on the interaction area.

The identification of such a layer has proved extremely difficult, as at the end of the burning stage (laser radiation turned off) the limitation of oxygen access to the burning reaction is suppressed and the nitride is transformed into an oxide [268] as a result of further cooling in air of the zirconium target, so that at the end of the process the target is covered only with a ZrO_2 layer.

No solution to the problem was available until a method developed in laser thermochemistry [204] was found suitable. It consists of keeping the thickness and content of the formed compounds unchanged until a certain final heating temperature, T_f, is reached, by blowing an inert gas jet onto the target.† A golden-yellow film was then noticed on the irradiated surface at $T \geqslant 1200$ °C (a characteristic of zirconium nitride and oxynitride).

The formation, under the specified conditions, of certain nitrogen-containing zirconium compounds has also been confirmed by the results of microhardness measurements and by x-ray and electron diffraction analysis of the interaction surface. Note that using the same diagnostic methods, the identification of

† We mention that the inert gas blowing method can also be successfully applied in conventional thermochemical studies. There also exists another method of tempering based upon thermoconductive heat removal into the bulk of a massive sample, which was applied in reference [268], when nitride coatings were obtained by scanning the surface of samples prepared from Zr and Hf with a cw CO_2 laser beam of 1–2 kW.

titanium nitrides or oxynitrides has not been possible in the laser burning conditions described above.

It should also be mentioned that a certain influence of the nitridation reaction upon the absorptivity $A(t)$ of the titanium sample laser burned in air was also observed, but titanium oxynitrides were found in very small amounts. In this case the main effect upon the quantity $A(t)$ during the burning stage comes from the conversion of TiO_2 to TiO [269]—an oxide which also exhibits metallic optical characteristics. In turn, the mechanism proposed in studies of titanium samples, concerning the variation of absorptivity as a result of the departure of the oxide from stoichiometry, must of course also act in the initial stage of zirconium ignition.

Any elaborate calculations and even simple numerical estimations concerning the absorptivity behaviour of zirconium samples laser-burned in air do not appear to be possible at this stage, not only due to lack of the required physical constants, but also, in general, because an adequate understanding of the observed processes is also lacking. Progress in this area is, however, very fast.

In conclusion, it has been shown, at least in the specific case of laser burning of zirconium in air, that methods of investigation of chemical kinetics using lasers can be successfully used not only in order to unveil the fundamentals of laser thermal processing of metals, but also to understand the nature of surface thermochemical reactions. Note that in this case the laser radiation performs two simultaneous functions, namely that of a clean source for high-temperature metal heating, and that of determining the optical properties of the metal–oxide system—properties that would continually change during the laser heating process.

Another mechanism that leads to the anomalous $A(T)$ behaviour during high-temperature metal burning in an oxidising medium has been theoretically studied in reference [253]. It starts from the known fact that in the interaction of a metal with a gas that contains oxygen a solid oxygen solution is formed in the metal, while an oxide grows on its surface. The oxygen flux from the oxide into the solution can considerably influence, in time, the oxide layer thickness, x. In the example of zirconium burning it has been shown [253] that during the limitation of the access of the gaseous phase to the reaction, a process of oxide dissolution into the metal can take place, soon followed by the oxide's regeneration. Obviously, when $x \to 0$, a significant decrease in the absorptivity A is to be expected.

It is important to note that in reference [253] the possibility of nitride compound formation was not taken into account in calculations. That is why to verify the idea experimentally one must use mixtures of O_2 and inert gases. Such experiments in O_2/Xe or O_2/Ar atmospheres confirmed the mechanism accounting for the 'anomalous' behaviour of the absorptivity of Ti and Zr targets, based on dissolution of the oxide layer during metal burning [269, 270].

7.2. Dynamic laser methods used to determine oxide optical and thermodiffusive constants

For the theoretical analysis of laser thermochemical heating of metal targets—even in the simplest case of a single-layer burnt area with a stoichiometrical content—it is necessary to have available a whole set of physical constants concerning: (i) power of thermal losses; (ii) oxide optical properties; (iii) chemical reaction rates.

Unfortunately not all these data are to be readily found in tables and research papers. Besides, it is necessary that the peculiarities of metal heating and oxidation under radiation action be properly taken into account. For instance, the optical constants of the oxide films that are generated on irradiated metal targets may differ from the optical constants of the metallic oxides prepared by other methods (as the oxide film lattice matches the surface of the metal foil lattice on which it is forming). There is also a probable influence of impurities and superficial defects, of the wavelength of radiation and so on. Accurate determination of the thermodiffusive constants represents in itself a difficult task. That is why even the earliest papers concerning oxidation of metals under the action of radiation questioned to what extent tabulated data, obtained via conventional methods, could be used in this case.

Today, the solution to some of these problems is possible, based on a comprehensive range of measuring methods proposed in papers on laser thermochemistry. Moreover, most of these methods provide for the experimental determination of the constants during laser irradiation—a fact which qualifies them as dynamic methods.

Let us deal for the moment with the constants of convective heat exchange, η, and of target blackening degree (emissivity), σ_0, on the basis of which the calculation of the thermal losses with thermally thin, heat-insulated foils is performed. The main difficulty met in this case derives from the constraint to consider the dependence on temperature $\sigma_0(T)$, which in turn is determined by the oxide layer thickness and its optical properties. For the calculations, averaged values of η and σ_0, obtained from the processing of experimental curves recorded during target cooling (occurring after the laser is switched off) are usually employed, with formula (6.15) for $Q_T \simeq 0$. The constants established in this way for copper foils, and used in papers [208, 211, 241, 242], were $\sigma_0 = 0.415$ and $\eta \simeq (2\text{–}5) \times 10^{-3}$ W cm^{-2} K^{-1}. We mention that the convective loss constant η calculated in this way is almost an order of magnitude higher than the one recommended in reference [2], $\eta = 5 \times 10^{-4}$ W cm^{-2} K^{-1}, which is an effect of the significant heat leaks through thermocouples in the case of laser measurements discussed above (even if the wire diameter was only $\simeq 0.1$ mm).

The method for determining the oxide optical constants, based on the analysis of the target absorptivity/reflectivity interferential oscillations is simple. As shown in reference [241], for relatively transparent oxide layers

($\alpha x \ll 1$), n and k are obtained according to absorptivity values in the first interferential minimum point, A^1_{min}, and maximum point, A^1_{max}, respectively

$$n = \left(\frac{A_{max}}{A_0 + (A_{min} - A_0) z/\pi} \right)^{1/2} \qquad k = \frac{A_{min} - A_0}{4\pi} n^2 \tag{7.8}$$

where z is determined from equation (6.29). In equation (7.8) the assumption $A_M(T) \simeq A_0$ is made. To illustrate, let us quote the determinations performed on zirconium and copper targets [241, 252] that allowed for the calculation of certain values $n = 1.35$–1.4, $k = 0.05$–0.08 in the case of ZrO_2 and $n = 1.1$–2.5 and $k \simeq 0.02$ for the copper protoxide, Cu_2O.

So far there is no strong evidence to back up the assumption that significant differences would exist between the values of the optical constants of the oxide layers forming on metals through laser heating, and the values n and k for single crystals of the respective oxides obtained by conventional methods. When such differences were noticed, they were either due to certain measurement errors or to the fact that a series of important factors mentioned in papers [260, 271] were ignored; these concern both the diffusion of the laser radiation incident on the rough surface of the oxide layer, and different optical inhomogeneities occurring within the oxide, as an effect of differences in the growth rates of the oxide at the irradiated area.

In paper [267] the kinetic law for oxide growth in the high-temperature laser air-oxidation stage of a number of metals has been experimentally established. To this purpose, standard weighing techniques were used. The samples were irradiated in air for different times, t, at the same level of incident laser power, P. The experiment was designed in such a way that the total duration of irradiation exceeds significantly the time required for a stationary temperature to be established in the sample. This means that the reaction conditions were close to the isothermal ones. The growth in the sample mass, Δm, was determined for different irradiation periods, t. It was demonstrated that, with copper, nickel and vanadium, at $T \sim 1000\ ^\circ C$, the function $\Delta m(t)$ is almost parabolic, $(\Delta m)^2 \simeq K_2(T)t$ and so equation (6.2) can be used for processing experimental data. For titanium and cobalt, the kinetic curves present, in the case of small thicknesses of the oxide film, a region of linear oxidation, $\Delta m = K_1(T)t$, while for greater thicknesses of the interaction area the reaction evolves according to a parabolic law. Finally, for tungsten and molybdenum, the kinetic dependences at $T \sim 1000\ ^\circ C$ were almost paralinear. All these results are in good agreement with the data obtained from conventional (non-laser) investigations (see, for example, [197–201]).

Knowledge of the kinetic law also permits the determination of the corresponding kinetic constants.

Analysis of interferential phenomena also substantiates the dynamic methods for the determination of oxide thermodiffusive constants [224, 241]. Later we shall thoroughly study these methods, as they represent a suggestive and sufficiently general approach to the solution of laser thermochemical problems.

If one neglects thermal losses out of the heated blades ($Q = 0$), one can obtain with equations (6.2) and (6.9) a direct connection between the thickness of the oxide layer and the sample temperature

$$\frac{dT}{dx} = \frac{Px}{mcd_0} A(x) \exp\left(\frac{T_d}{T}\right). \tag{7.9}$$

Through the integration of equation (7.9) one obtains

$$\int_{T_0}^{T_f} \exp\left(-\frac{T_d}{T}\right) dT = \frac{P}{mcd_0} \int_{x_0}^{x_f} xA(x)\, dx \tag{7.10}$$

where T_0, x_0 and T_f, x_f are the initial and final values of the sample temperature and oxide layer thickness, respectively.

By determining the temperatures T_{1m} and T_{2m} that correspond to the same minimum in the absorptivity, $x_f = x_{min}$, at different values of radiation power, P_1 and P_2, and of the target mass, m_1, m_2, the value of T_d can be found. Indeed, assuming $x_0 = 0$ and taking into account that, as a rule, $T \ll T_d$ and $T - T_0 \gg TT_0/T_d$, one gets the relations

$$T_d = \frac{T_{1m} T_{2m}}{T_{2m} - T_{1m}} \ln\left[\frac{P_2 m_1}{P_1 m_2}\left(\frac{T_{1m}}{T_{2m}}\right)^2\right] \tag{7.11}$$

from which the activation temperature of diffusion can be determined proceeding from experimental values. With T_d known, d_0 can then be determined, by means of equation (7.10), because the integral in the right-hand side of this relation is relatively easy to evaluate in the point A_{min}.

By means of this method (figure 7.6), the activation temperature of the copper oxidation reaction, $T_d \simeq 10\,000$ K, has been determined [241]. Let us note that this value of T_d differs considerably from the Wagner constant T_{dW} for Cu_2O (in isothermal conditions $T_{dW} \simeq 18\,800$ K [197, 198]). The quantity d_0 has not been measured in [241], being used as a fitting parameter when calculating the heating curves. The best agreement with the experimental records, corresponding to a diffusion activation temperature $T_d \simeq 10\,000$ K, has been obtained for $d_0 \simeq (1–3) \times 10^3$ cm^2 s^{-1}, again differing considerably from the corresponding Wagner value, $d_{0W} \approx 0.299$ cm^2 s^{-1}.

The experimental evolution $A(T)$ and the corresponding curve, calculated numerically using Cu_2O optical and thermodynamical constants, obtained by means of the above-described dynamical laser methods, can be followed in figure 6.20. Unfortunately the calculations of temperature evolution, $T(t)$, gave values of the activation stage duration, t_a, and time of melting, t_m, several times lower than the experimental ones.

This discrepancy derives from the following: when establishing equations (7.9)–(7.11), thermal losses, which become significant at levels of incident

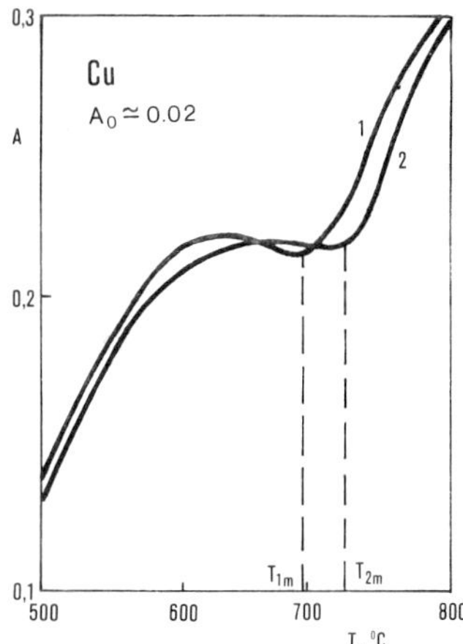

Figure 7.6 Experimental dependences $A(T)$, for two different levels of the incident laser radiation power, $A_0 = 0.02$. (1) $m_1 = 50$ mg, $P_1 = 15$ W, $T_{1m} = 693$ °C; (2) $m_2 = 50$ mg, $P_2 = 21$ W, $T_{2m} = 718$ °C.

laser power, P, and of sample mass, m, comparable with those being used in [241], have not been taken into account. This can be proved by integrating equation (6.9) and using (6.2), between the limits $x_0 = 0$ at x_m

$$\frac{1}{P}\int_0^{T_m} \exp\left(-\frac{T_d}{T}\right) dT + \frac{1}{Pmcd_0}\int_0^{x_m} xQ(x)\, dx = \frac{1}{mcd_0}\int_0^{x_m} xA(x)\, dx. \tag{7.12}$$

Calculations show that the second term on the left-hand side of equation (7.12), determined by the thermal losses—which is the one neglected for the sake of simplicity in (7.10)—represents, under the experimental conditions from [241], about 30% of the first term at $P_1 \simeq 21$ W, and as much as 50% of it at $P_2 \simeq 15$ W, and therefore begins to play a significant part.

Accordingly, an improved method of determination of T_d and d_0 has been proposed in [224]. For two different values of radiation power P_1 and P_2, in the case of the two samples of comparable masses, $m_1 \simeq m_2 \simeq m$, one gets, by means of equation (7.12)

$$\frac{T_d}{mcd_0 T_{2m}^2}\int_0^{x_m} x\left(\frac{P_2}{P_1}Q_1 - Q_2\right) dx = \exp\left(-\frac{T_d}{T_{2m}}\right) - \frac{P_2 T_{1m}^2}{P_1 T_{2m}^2}\exp\left(-\frac{T_d}{T_{1m}}\right). \tag{7.13}$$

The dependences $y_1 = xQ_1$, $y_2 = xQ_2$ and $y_3 = x(PQ_2/Q_1 - Q_2)$, derived from the corresponding target cooling curves, are given in figure 7.7.

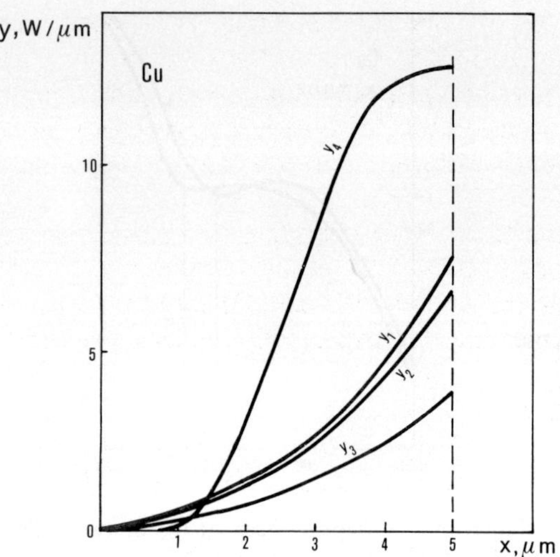

Figure 7.7 Evolution of the parameters y_1, y_2, y_3, y_4 as functions of the oxide layer thickness, x.

In order to obtain a full set of equations in relation to d_0 and T_d, one adds to (7.13) the relation

$$d_0 \exp\left(-\frac{T_d}{T_{2m}}\right) = \frac{T_d}{mc_0 T_{2m}^2} \int_0^{x_m} x(P_2 A - Q_2)\,dx \tag{7.14}$$

which results from (7.12). We specify that the product on the left-hand side of equation (7.14) is just the rate constant of the chemical reaction at the temperature T_{2m}. Eliminating the constant d_0 from (7.13) and (7.14) we obtain

$$T_d = \frac{T_{1m} T_{2m}}{T_{2m} - T_{1m}} \ln\left(\frac{P_2 T_{1m}^2}{P_1 T_{2m}^2} \frac{1}{1-\xi}\right) \tag{7.15}$$

where

$$\xi = \frac{\displaystyle\int_0^{x_m} x(P_2 Q_1/P_1 - Q_2)\,dx}{\displaystyle\int_0^{x_m} x(P_2 A - Q_2)\,dx}. \tag{7.16}$$

Using (6.5) and the dependence $y_2(x)$, one can plot the function $y_4 = x(P_2 A - Q_2)$ and determine the value $\xi = 0.112$ for the experimental conditions in reference [241].

In this way one finds, by means of equation (7.15), $T_d \simeq 15\,500$ K in the case of the oxide Cu_2O, and accordingly, from equation (7.14), $d_0 \simeq 1.4\ \mathrm{cm^2\,s^{-1}}$.

The results of the calculations concerning the temperature evolution $T(t)$ are presented in figure 7.8, based on these values of T_d and d_0, using the experimental values of the constants n, k, η, σ_0 and taking into account the dependence on temperature of copper's absorptivity. A good agreement between the results of these calculations and the experimental data can be noticed.

It is interesting to compare the rate of growth of the oxide film forming on copper during the laser heating process $(dx/dt)_L$ with that obtained through calculations based on Wagner's theory $(dx/dt)_W$. Starting from the values T_d and d_0 measured in reference [224] for a temperature $T \simeq T_{2m} \simeq 991$ K, one gets

$$\frac{(dx/dt)_L}{(dx/dt)_W} = \frac{d_0}{d_{0W}} \exp\left(\frac{T_{dW} - T_d}{T}\right). \tag{7.17}$$

We mention that the relation $(dx/dt)_L/(dx/dt)_W \gg 1$ is also confirmed by another, simpler, analysis. In fact, let us integrate equation (6.2) in relation to the time lag between the moment of observation of the first maximum, t^1_{max}, and that corresponding to the first minimum value, t^1_{min}, using the experimentally obtained values t^1_{max} and t^1_{min} at a fixed temperature $T \simeq T_{2m}$. Although in this way an overestimation of the oxide layer thickness, x, is made, the calculations based on the Wagner constants d_{0W} and T_{dW} give variations of the oxide layer thickness $\Delta x \sim 10^{-5}$–10^{-6} cm, while according to interferential oscillation analysis, the actual variation of the oxide layer thickness cannot be less than $\simeq \lambda/4\eta \sim 1$ μm.

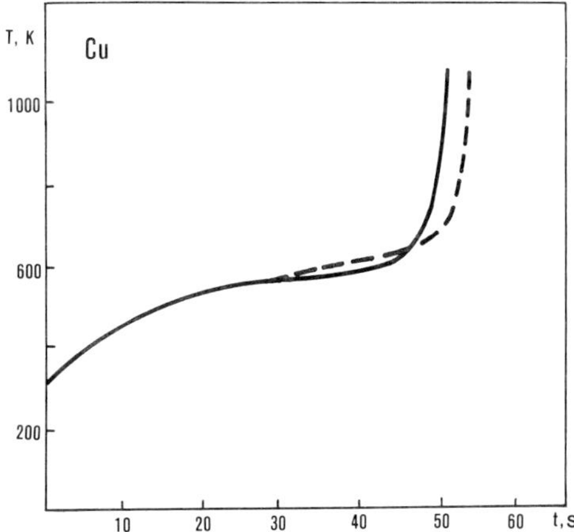

Figure 7.8 Time evolution of the temperature of a copper sample, $T(t)$, during laser heating: experimental (full curve) and calculated (broken curve).

Thus, from the data reported in references [224, 241] for laser oxidation of copper in the examined range of temperatures and heating rates, the growth rate of the oxide layer is considerably higher than in isothermal conditions characterising the chemical reaction. It is possible that this effect is determined by the acceleration in the diffusion of the reaction components through the interstices between crystallites and through microcracks, which, as electron microscopy investigations [53] have proved, occur during fast laser heating of oxidising metal samples. The possibility of oxide cracking under these circumstances results also from the calculations reported in reference [272]. Besides, by laser heating, a gradient of temperature is always apparent in the oxide layer $|\Delta T| \sim A\bar{I}/k_{ox}$ (k_{ox} is the thermal conductivity of the oxide). As has already been shown [273], in this instance an additional term occurs in the expression of the diffusion flux, proportional to the coefficient of the thermoelectromotive force in the oxide. It is to be expected that in the case of copper oxides this effect would also lead to the acceleration of the oxidation reaction.

A similar result, expressed by the relation found, namely $(\mathrm{d}x/\mathrm{d}t)_L/(\mathrm{d}x/\mathrm{d}t)_W \gg 1$, was obtained [252] for the stage of laser activation of zirconium oxidation in air. On the other hand, the method of determination for d_0 and T_d proposed in reference [231] for the growth stage of the oxide film before the oxidation activation, at T constant ($AP = Q$), has resulted, for titanium, in a good agreement with data in the literature.

The kinetics of laser oxidation of titanium targets has been the most thoroughly studied, by several teams in parallel, using different methods [211, 231, 249, 274].

In figure 7.9 are shown the dependences of the oxidation rate of titanium $v_{ox} = d_0 \exp(-T_d/T)$, as a function of the inverse temperature, obtained in various papers. One may see that the results are markedly different from one another when compared with the values v_{ox} obtained by non-laser measurements [199, 202, 265] corresponding to the same temperature range and to identical heating rates ($T \geqslant 1200$ K, $\mathrm{d}T/\mathrm{d}t \leqslant 300$ K s^{-1}). For instance, non-laser methods yield, at a temperature $T \simeq 1400$ K, values of v_{ox} an order of magnitude greater than those obtained in [214], and three orders of magnitude greater than those obtained in reference [249] by laser techniques.

According to the analysis carried out in reference [252], the abnormally low values of the oxidation rate, v_{ox} measured in [249], do not originate from a laser specificity of the titanium oxidation process, but are determined by a systematic error. On the other hand, the main error source when determining v_{ox} [214] lies in the non-equilibrium oxidation of the target surface as a result of non-uniform target illumination and of temperature gradients. Taking into consideration this factor, the theoretical processing of the experimental data resulted in a good agreement between v_{ox} and non-laser measurement data. The oxidation constants obtained in this way, $d_0 \simeq 330$ m^2 s^{-1} and $T_d \simeq 33\,000$ K, have been successfully used when

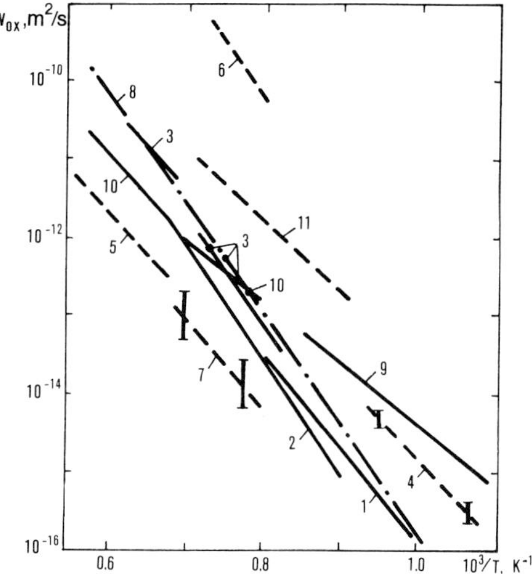

Figure 7.9 Rate constants in the case of parabolic oxidation: Titanium: (1) [202], (2) [220], (3) [215], (4) [201], (5) [213], (6) [206], (7) [227], (8): fitting of experimental data. Zirconium: (9) [202], (10) [215], (11) [252].

calculating heating and oxidation curves of titanium targets under the action of cw CO_2 laser radiation (figure 7.3).

Let us emphasise that the sensitivity to temperature inhomogeneities is characteristic of all thermographical methods, and the allowed temperature drop, determined by the inequality $\Delta T \ll T^2/T_d$, should not in practice exceed 10 K. This requirement is adequately met only by the experiments with thermally thin and freely supported (on thermocouples) copper foils, whose surface is uniformly illuminated. Even in this case a number of other factors causing instabilities in the growth of the oxide layer [275–282] have to be taken into account—a fact that is also true for the temperature drop in the oxide layer and not only in the foil [275, 283].

At the end of this section, we mention two interesting effects that may occur with the development of different instabilities in the process of laser oxidation of metals. First, we point to the synthesis of huge Cu_2O [284] and V_2O_5 single crystals with large surfaces [285]. Secondly, in [263], for laser heating of vanadium, significant electromotive forces have been noticed in the metal–oxide system, the system being affected by complex dynamic transformations. These have been ascribed to oscillations in the spatial temperature distribution and to the non-stoichiometry of the oxides.

One more very important point is in order: so far we have considered only the action of infrared radiation. The situation may become much more complicated when visible and especially ultraviolet radiation is used. In this case a series of data is available, showing that the laser does not act as a simple thermal source, and that the oxidation rate may be much higher than for the same time–temperature history in traditional reaction conditions and even in comparison with CO_2 laser heating. Such peculiarities could be related to the generation of electron–hole pairs near the semiconductor (oxide) surface, and the photoemission of electrons.

Chapter 8 The Choice of Optical Irradiation Conditions

In this concluding chapter, several examples are given of the selection of irradiation conditions in such a way that the laser heating of metals is obtained with the minimum expenditure of energy and/or minimum irradiation times. The results illustrate the philosophy of the book—that a proper choice of truly optical irradiation parameters is possible only on the basis of a detailed account of the physical and chemical processess occurring on surfaces exposed to laser irradiation.

Efficient implementation of laser technology requires that the irradiation conditions be chosen such that the heating of the sample up to the desired temperature be obtained with minimum energy expenditure and/or minimum irradiation times (see, for example, [286–291]).

The proper choice of optimal irradiation parameters is possible only on the basis of a detailed account of different physical and chemical processes which occur on the laser-heated surface. The physical parameters concerned are the temperature-dependent optical and thermophysical constants of the metals, the irradiation spot size and target geometry, the time behaviour of the laser beam power, its intensity, polarisation, angle of incidence, etc. In the case of a thermochemical interaction process some additional gain can be obtained by the increase (within certain limits) in the oxygen partial pressure or blowing a jet of air or oxygen on the irradiation spot.

We shall discuss in this chapter several examples of the choice of optimal irradiation parameters, both for the physical and thermochemical mechanisms of laser-induced heating of metals. In all cases the approach to the calculations is based on the account of the two competing processes: the absorption of radiation energy and the removal of heat from the irradiation spot by thermal conductivity, convection and light emission.

8.1. Sample heating in chemically inert gases

8.1.1. Choice of laser radiation energy and pulse shape

We shall discuss this case first on the assumption of a constant absorptivity, and then by taking into account the temperature variation of the absorptivity during laser irradiation [290, 291].

(i) $A(T)=\text{const}$. Leaving aside the heat exchange with the environment and using the approximation of a semi-infinite solid, it was shown that the incident fluence required to heat a metal sample up to a temperature $T_f \leqslant T_m$, and to a depth z_f, under the action of a pulse of rectangular time-shape, can be minimised ($E_s = E_s^{min}$) by selecting the following optimum values for the incident laser radiation intensity, $I_0 = I_{opt}$, and the pulse duration $\tau_p = \tau_p^{opt}$:

$$I_{opt} \simeq \frac{1.85 k_T T_f}{A z_f} \qquad \tau_p^{opt} \simeq \frac{1.33 z_f^2}{\kappa}. \tag{8.1}$$

The corresponding minimum of the laser fluence is

$$E_s^{min} \simeq 2.47 E_s^z \qquad \text{where } E_s^z = c\rho z_f T_f \tag{8.2}$$

with c, ρ, k_T, κ the specific heat, the density, the thermal conductivity and the thermal diffusivity of the metal, respectively.

We mention that according to the equation given in reference [292]

$$\frac{E_s}{E_s^z} = \frac{\sqrt{\pi}}{4\gamma[\exp(-\gamma^2) - \sqrt{\pi}\kappa \operatorname{erfc} \kappa]} \tag{8.3}$$

where $\gamma = z_f / 2\sqrt{\kappa \tau_p}$, the ratio E_s/E_s^z can markedly depart from the optimum, i.e. for $\gamma \gg \gamma_{opt}$ or $\gamma \ll \gamma_{opt}$, with $\gamma_{opt} = 0.433$, the ratio on the left-hand side of equation (8.3) can be an order of magnitude larger than the minimum value E_s^{min}/E_s^z.

A large gain in comparison with the energy required to heat a metal layer is also to be obtained when applying a laser δ pulse

$$I(t) = B\delta(t). \tag{8.4}$$

For a semi-infinite sample we have

$$E_s^{min} \simeq 2.06 E_s^z. \tag{8.5}$$

(ii) $A(T) = A_0 + A_1 T$. When one takes into account the temperature dependence of the absorptivity and the convective thermal losses, the heating of a thermally thin foil is described by the following law

$$T(t) = \frac{A_0 P}{A_1 P - 2\eta S_s}\left[\exp\left(\frac{A_1 P - 2\eta S_s}{mc} t\right) - 1\right] \tag{8.6}$$

where $2S_s$ is the total surface of the exposed and opposite surfaces of the foil, and η is the constant of convective heat exchange.

The energy required to heat the sample up to a final temperature T_f depends on the power of the incident radiation, according to the relation

$$E(P) = \frac{mcT_f}{A_0}\frac{\ln(1+y)}{y} \qquad y = \frac{A_1 P - 2\eta S_s}{A_0 P} T_f. \tag{8.7}$$

From equation (8.7) we infer that the minimum energy—and not fluence as in the case of a semi-infinite sample—corresponds in this case to $P \to \infty$, for a δ time-shaped heat source, and equals

$$E_0^{\min} = \frac{mc}{A_1}\ln\left(1 + \frac{A_1 T_f}{A_0}\right). \tag{8.8}$$

Consequently, when the $A(T)$ dependence is taken into account, a reduction of the required energy can be obtained by shortening the laser pulse and, correspondingly, by increasing the radiation power. Of course, care must be exercised not to exceed the thresholds beyond which other processes are triggered onto the irradiated surface, as for instance intense vaporisation and plasma generation. For the case in discussion, the existence of an optimum regime is related to competition between two factors: the convective thermal losses, and the increase in the sample's absorptivity with temperature. Obviously, the optimum regime is reached when thermal losses are reduced to zero.

8.1.2. Optimum heating with a source in motion

We shall first introduce some simple, straightforward evaluations for the case of small irradiation spots on the surface of semi-infinite samples, which the laser beam scans with a velocity v.

The irradiation time τ_v with the scanning cw laser beam—which is equivalent to the duration τ_p of the laser pulse in the case of laser irradiation of a fixed zone of the sample surface—results from the ratio

$$\tau_v = \frac{2R_s}{v}. \tag{8.9}$$

For efficient laser heating of a surface layer of the sample, the heat wavelength (thermal diffusion depth), $l_{th} = (\kappa\tau_v)^{1/2}$, has to be kept much smaller than the radius of the irradiation spot—for in this case the thermal losses due to the thermoconduction mechanism would indeed be very small. Accordingly one infers the condition

$$\left(\kappa\frac{2R_s}{v}\right)^{1/2} \ll R_s \tag{8.10}$$

or

$$v \gg \frac{2\kappa}{R_s}. \tag{8.11}$$

According to equation (8.11), we obtain for $R_s = 1$ mm and metals such as titanium and steel ($\kappa \simeq 10^{-1}$ cm^2 s^{-1}), a velocity $v \gg 2$ cm s^{-1}. For smaller irradiation spots and/or metals exhibiting much higher thermoconduction (Cu, Al, etc) one can use, according to equation (8.11), even larger scanning velocities on the metal surface.

In order to evaluate the lower limit of the scanning velocity required to heat a surface layer of metal of thickness $z_f \ll R_s$, one has to substitute z_f for R_s in the right-hand side of equation (8.10).

The upper limit for the scanning velocity v follows from the limitation to be applied to the incident laser power—since, along with the increase in velocity v, in order to provide for the required temperature inside the surface layer one has to correspondingly increase the incident laser radiation power per unit area of the irradiated surface.

In this respect we shall introduce in the following several useful formulae obtained in reference [291], in the limits of high and low displacement velocities of the source, and for a Gaussian distribution of energy over the surface of the foil, of thickness h

$$r_1 = \frac{vz_f^2}{2\kappa} \qquad T_f = \left(\frac{2}{e}\right)^{1/2} \frac{A\kappa I_0 R_s}{k_T z_f}$$

$$\text{for } z_f \ll h \qquad \frac{vR_s}{2\kappa} \gg 1 \qquad \frac{vz_f^2}{2\kappa R_s} \gg 1 \tag{8.12}$$

$$r_1 = 0.54R_s \qquad T_f = 2.16\left(\frac{\kappa R_s}{\pi v}\right)^{1/2} \frac{AI_0}{k_T}$$

$$\text{for } z_f = 0 \qquad \frac{vR_s}{2\kappa} \gg 1 \tag{8.13}$$

$$r_1 = \frac{vR_s^2}{4\kappa} \ln \frac{16\kappa^2}{\gamma_E e^2 v^2 R_s^2} \qquad T_f = \frac{AI_0 R_s^2}{4k_T h} \ln \frac{16\kappa^2}{\gamma_E v^2 R_s^2}$$

$$\text{for } \frac{vR_s}{2\kappa} \ll 1 \tag{8.14}$$

where $\gamma_E = 1.781$ is the Euler constant. The temperature is maximum, and equals T_f at the point $r = r_1$ (and not at $r = 0$), in the coordinate system related to the moving source and directed along its displacement.

The energy expenditure for the processing of a unit length of material results from [291]

$$E_v = \frac{\pi R_s^2 I_0}{v} \tag{8.15}$$

which becomes minimum for $v \to \infty$

$$E_{\mathrm{s}}^{\min} = \left(\frac{\pi^2 e}{8}\right)^{1/2} E_{\mathrm{v}}^{z} \qquad E_{\mathrm{v}}^{z} = 2R_{\mathrm{s}} z_{\mathrm{f}} c\rho T_{\mathrm{f}}. \tag{8.16}$$

8.2. Ways to improve the efficiency of thermochemical metal processing

8.2.1. A proper selection of the cw laser radiation power

The energy expenditure for heating the target under the action of a radiation source of constant power, P, up to the melting temperature, T_{m}, is $E_{\mathrm{m}} = Pt_{\mathrm{m}}$. Here t_{m} stands for the melting initiation time. In figure 8.1, the theoretical dependences $E_{\mathrm{m}} = E_{\mathrm{m}}(P, A_0)$ for a copper foil heated through cw CO_2 laser irradiation are plotted [292]. The curves exhibit a minimum for a certain incident power, P_{opt} (the irradiated spot area was identical in all cases, $S_{\mathrm{s}} \simeq 0.1\ \mathrm{cm}^2$).

The existence of such a minimum is related to the fact that, for low values of the ratio P/S_{s}, most of the energy absorbed by the metal is compensated by thermal losses. Therefore, when A_0 increases, the position of P_{opt} is shifting towards lower values of incident laser power. On the other hand, at high levels of incident radiation power the heating rate is too high, and in the time lapse t_{m} the oxide film does not grow significantly. Therefore, at $P = P_{\mathrm{opt}}$, the most favourable ratio between the increase in the absorptivity A as an effect of the metal oxidation, and thermal losses, is obtained.

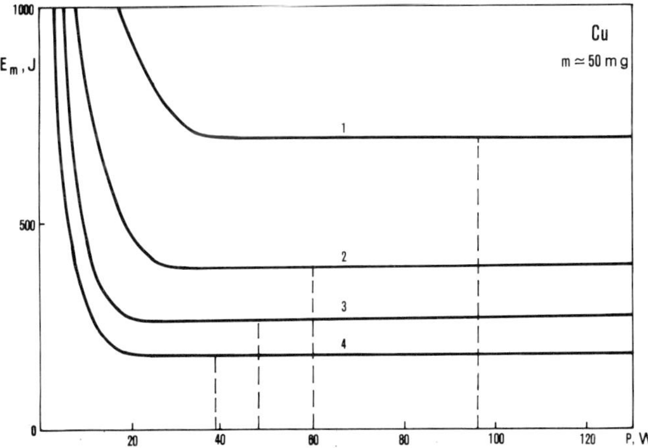

Figure 8.1 Dependence of the energy expenditure for the laser-induced heating up to the melting point of copper samples, as a function of the incident radiation power (1) $A_0 = 0.02$, (2) $A_0 = 0.04$, (3) $A_0 = 0.06$, (4) $A_0 = 0.1$. The broken lines indicate the positions of the minimum values on the curves $E(P)$; $m \simeq 50$ mg.

Also for $P > P_{opt}$, E_m exhibits only a weak variation and in order to be within the range of P values which are energetically favourable we have only to pass fast enough through the low-power range, and be careful not to exceed P_{opt} by too much.

It is also important to evaluate the advantages of the use, at a given laser radiation intensity, of an oxidising atmosphere (air or other oxygen-containing mixtures), as against an inert atmosphere (nitrogen or argon for instance). In both cases the thermal power loss is almost the same, while in air the absorptivity, A, increases to a higher degree during the laser heating process. Figure 8.2 shows the dependence of $E_m^{air}/E_m^{N_2}$, the ratio between the energies necessary for heating in air/nitrogen up to T_m (at $P = P_{opt}$), on the initial absorptivity of some copper targets. This curve demonstrates in a convincing manner the advantages of a thermochemical laser metal heating regime and is in good agreement with experimental results [237].

An example of optimised control of the oxidation reaction, aiming at enhancing the efficiency of the thermal processes involved in laser irradiation of metals, consists in the use of two lasers, a pulsed and a cw one, with beams directed onto the same region of the target surface [245, 246, 256, 257]. Thus, one or more laser pulses (the pulsed laser can operate on a wavelength other than 10.6 μm) can provide for a more efficient heating of an oxidising target (steel, titanium) under the action of a cw CO_2 laser source.

In fact, the pulsed irradiation caused an increased sample absorptivity as against A_0, and a shortening of the heating activation stage. Moreover, the pulsed preheating of the surface can cause target ignition, so that, by blowing a gas jet upon the target, continuous metal burning can be ensured.

Theoretical analysis of the thermochemical problem has indicated [293] that the best profit—from the energy viewpoint—for laser heating of an oxidising metal can be obtained by applying two short auxiliary pulses. The first, at the very onset of cw laser irradiation, performs the fast initiation of the oxidation reaction, while the second pulse, intervening after a certain

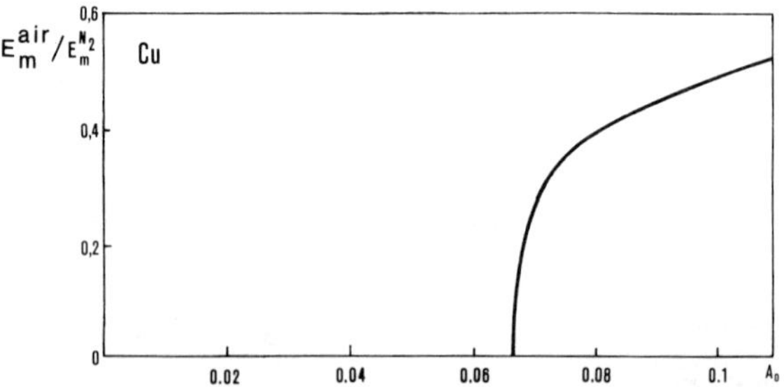

Figure 8.2 Evolution of the energy gain $E_m^{air}/E_m^{N_2}$, corresponding to the laser heating up to the melting point, T_m, resulting from the activation of the reaction of surface oxidation, as a function of the initial absorptivity of copper samples.

while, compensates for thermal losses that are rapidly growing in the final stage of high-temperature heating by the main laser beam.

8.2.2. *High repetition rate* (HRR) *pulsed irradiation regime*

The oxidation of metals under multipulse laser irradiation has also been investigated [294]. A CO_2 laser, generating pulses of an energy $E_0 = 0.25$ J, an average duration of $\tau_p \simeq 10$ μs and at a variable repetition frequency rate $f \leqslant 300$ Hz, was used.

The calorimetric method for recording the average values $T(t)$ and $A(t)$ was no different from that used for the metal foil heating under the action of CW CO_2 laser radiation. Only the irradiation spot area onto the metallic surface, S_s, has been increased. It had in this case an approximately oval shape with characteristic dimensions of 4×3 mm^2 (as against an area of 7×4 mm^2 for the faces of the irradiated sample). This increase was necessary to prevent the low-threshold optical breakdown of air in front of the irradiated target surface.

In figure 8.3 we have plotted the evolutions of $T(t)$ and $A(t)$ for a copper target, irradiated by a multipulse laser at $f \simeq 150$ Hz, $A_0 = 0.045$ and $P \simeq E_0 f \simeq 20.6$ W. The first notable fact is that the oxidation activation is achieved in the high repetition rate pulsed irradiation regime at lower sample temperatures than in the case of CW laser heating (compare figures 8.3 and 6.14). With repeated irradiation, a higher number of $A(T)$ interferential oscillations is recorded, while the maximum absorptivity value reaches a level of 0.7–0.8 in the vicinity of the melting point.

All these elements prove that for the same value of the average power, growth of the oxide layer is induced more quickly with HRR pulsed irradiation than with CW irradiation.

A quantitative comparison between the heating efficiencies of metals subject to oxidation in the two regimes has been attempted. Unfortunately, at the

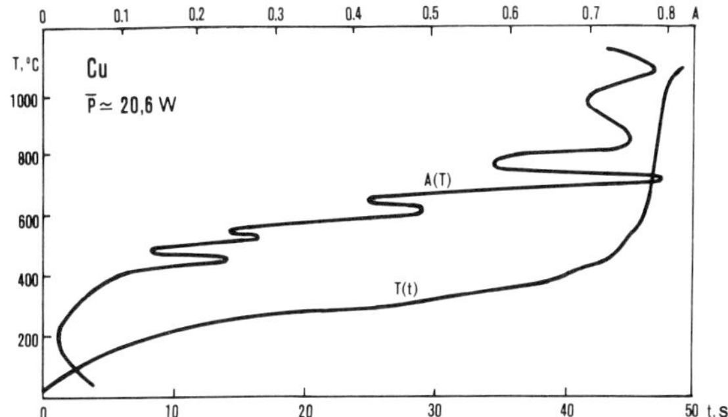

Figure 8.3 The characteristic heating curves for a copper sample heated in air by means of a periodically pulsed CO_2 laser ($f \simeq 150$ Hz, $m \simeq 75.7$ mg, $P \simeq 20.6$ W).

maximum average power used in experiments [294], $P_0 \simeq 30$ W, it proved impossible to induce target melting with cw irradiation onto a spot of $S_s \simeq 28$ mm^2, because of important thermal losses. Therefore, it was not possible to determine the gain in relation to the energy expenditure necessary to achieve target melting, E_m, that could be possible by passing from one irradiation regime to the other. Anyway, experimental data analysis concerning the melting time, t_m, and the oxidation activation time, t_a, respectively, as well as numerical calculations [294, 295], have indicated that this gain can result in reducing the energy expenditure up to five times, depending on the values of the irradiation parameters.

The increased efficiency of the thermochemical action of multipulse laser irradiation can be explained by the fact that during the action of every laser pulse, as well as after a certain time following its interruption, the target surface is overheated in comparison with the average temperature reached inside the sample. Simple calculations have shown that the value of this surface overheating, ΔT, may become significant at levels of pulsed laser radiation intensity, I_0, close to yet less than the breakdown plasma threshold. For example, even in the case of a highly reflecting metal having a high thermal conductivity (e.g. copper) for $A_0 \simeq 0.025$, $I_0 = 10^6$ W cm^{-2}, $\tau_p \approx 10$ μs, the surface overheating is $\Delta T \simeq 10$ °C, and, in the case of the formation of an oxide layer, it can reach as high as several hundred degrees. If we take into account the rapid (exponential) oxide growth dependence on temperature it is easy to see that, even with these levels of surface overheating, a marked intensification occurs in the heating rate of the oxidising metal samples. Besides, the pulsed thermal shock results in intense fracturing (cracking) of the oxide layer, which, by facilitating oxygen access into the depth of the metal, leads to an increase in the chemical reaction rate.

It has also been noticed [2, 296] that thermal metal (steel) processing through the action of periodically pulsed CO_2 lasers is several times more efficient than through the action of cw lasers. However, no connection has been shown between this effect and the thermochemical mechanism, the superior efficiency obtained with periodic pulse irradiation being considered to be a result of metal surface damage, and of the corresponding increase in A_0, by the repeated action of the power pulses.

8.2.3. Changing the angle of incidence of the radiation

There is another avenue—perhaps surprising at a first glance, though obvious enough after some thought—by which to modify the irradiation conditions so as to obtain substantial (in some cases by more than an order of magnitude) energy savings during the heating of an oxidising metal. It consists of oblique orientation of the laser radiation onto the target $\theta \neq 0°$ (oblique incidence) and the use of polarised radiation.

The effect was first noted [297] in the case of cw CO_2 laser irradiation of some tungsten targets ($P \simeq 700$ W). The targets were foils with a thickness

$h \simeq 200\ \mu m$. The irradiation spot diameter, at $\theta = 0°$, was $\simeq 5$ mm—considerably less than the samples' widths. The evolution of the ignition time as a function of the angle of incidence of the radiation $t_{ig}(\theta)$, at a constant incident laser power has been experimentally studied (figure 8.4).

It can be seen that over the entire range of θ as investigated, the value t_{ig} is, in the case of oblique radiation incidence, systematically lower than at normal incidence, even if one takes into account the increase in the area of the irradiated spot ($\sim 1/\cos\theta$). The curve $t_{ig}(\theta)$ has a minimum at θ_{min} (40–60 °C), the corresponding minimum value of $t_{ig}(\theta_{min})$ being almost 40 times lower than $t_{ig}(0)$.

We nevertheless notice that no special measurements of the degree of radiation polarisation were reported in [297]. In [231] control experiments were carried out, using HRR pulsed, polarised, CO_2 laser radiation. Titanium foils were used as targets ($h \simeq 50\ \mu m$, $R_s \simeq 3$ mm). The radiation could be polarised both in the incidence plane (index $\|$) and perpendicular to it (index $\perp$), which was obtained by rotating the laser tube windows placed at the Brewster angle.

The relative evolutions of the oxidation activation times of titanium targets in the two cases, $k_a^{\|} = t_a^{\|}(\theta)/t_a^{\|}(0)$ and $k_a^{\perp}(\theta) = t_a^{\perp}(\theta)/t_a^{\perp}(0)$, can be followed in figure 8.5. The gain in relation to the energy expenditure, obtained with oblique incidence of laser light, characterised by the value $1/k_a$, depends on the choice of the radiation polarisation direction. This gain is lower in the case of a titanium target as compared with tungsten (the value θ_{min} also differs).

The energy gain obtained by using obliquely incident radiation is determined by the angular dependence of the absorptivity of the oxide–metal system

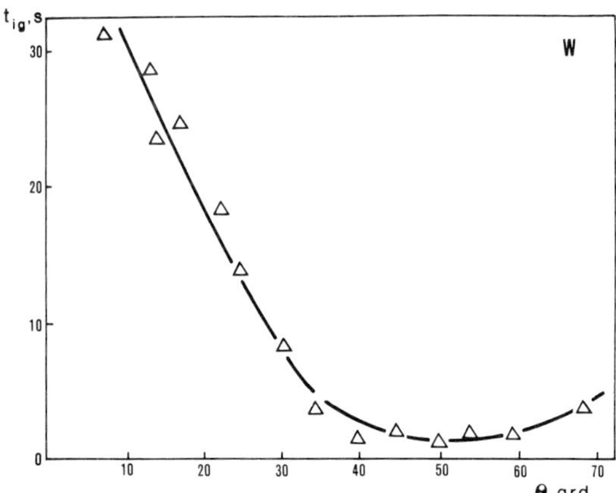

Figure 8.4 Experimental dependence of the ignition time, t_{ig}, of a tungsten foil, as a function of the incidence angle, θ, of cw CO_2 laser radiation.

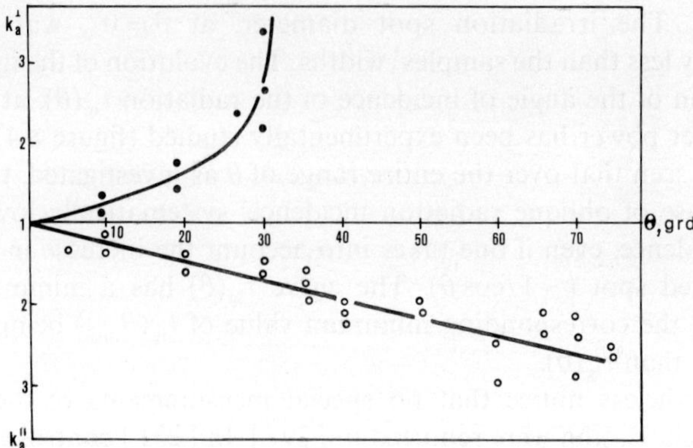

Figure 8.5 Variation of dimensionless parameters $k_a^\parallel$ and $k_a^\perp$ with the increase in the incidence angle, θ, of laser radiation polarised in the incidence plane ($\parallel$), and normal to it ($\perp$).

and by the strong interdependence between heating and oxidation processes. This problem has been studied theoretically and reported in reference [118].

It is known (see also §1.1.1.6), that the absorptivity of metal–oxide systems evolves, in relation to the value of the angle of incidence, in the following way: in the case of radiation polarised normal to the incidence surface, the absorptivity $A_\perp(\theta)$ decreases monotonically with the rise in θ; whereas in the case of radiation polarised in the incidence plane, the absorptivity $A_\parallel(\theta)$ has a peak corresponding to the radiation incidence under the Brewster angle. The corresponding curves calculated for Cu and Cu_2O are given in figure 8.6. More intricate calculations are required for the determination of the shape of the $A(\theta)$ curves for metal–oxide systems, as in this case A also depends on the oxide thickness x and the interference phenomena in the metal–oxide system must be taken into consideration. The results of such calculations for the system $Cu + Cu_2O$ ($x \simeq 3.5\ \mu m$) are also represented in figure 8.6. It is obvious that a significant gain in relation to the thermal losses can be obtained by means of radiation polarised in the incidence plane ($\parallel$), in good agreement with the experimental results [231].

Taking into account the great practical importance for metal processing technology, we shall deal more thoroughly with metal heating under the action of laser radiation polarised in the incidence plane. The following approximate expression for $A_\parallel(x, \theta)$ has been established [298], which represents the generalisation of equation (6.5) for the case $\theta \neq 0$

$$A_\parallel(x, \theta) = \cos\theta \, \frac{n^2 B(\theta)\cos\theta + (2kn/k_k)[\beta x - (1 - 2\sin^2\theta/n^2)\sin\beta x]}{n^2\cos^2\theta + [(k_k^2 - n^4\cos^2\theta)/n^2]\sin^2(\beta x/2)} \tag{8.17}$$

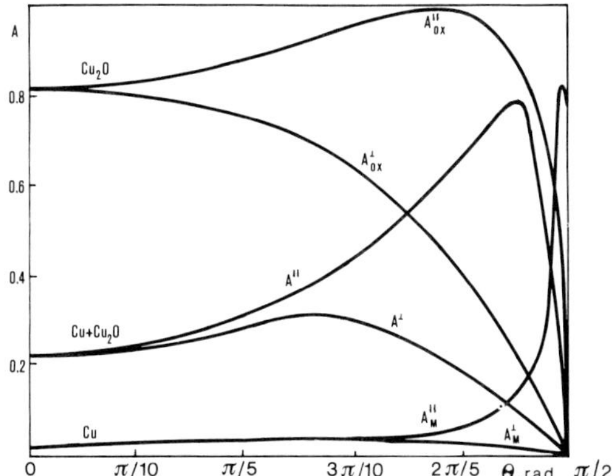

Figure 8.6 Calculated evolution of the copper 'cold' absorptivity (at room temperature), A_M, of the Cu_2O oxide, A_{ox}, and of the metal oxide system $Cu + Cu_2O$, A, as a function of the incidence angle, θ, of CO_2 laser radiation polarised in the incidence plane ($\|$), and perpendicular to it ($\perp$).

where $k_k = (n^2 - \sin^2\theta)^{1/2}$, $\beta = 4\pi k/\lambda$, $B(\theta) = (8A_0 \cos\theta)/[(A_0 + 2\cos\theta)^2 + 4\cos^2\theta]$. Formula (8.17) is valid for $A_0 \ll 1$, $k \ll 1$, $\alpha x \ll 1$ (where $\alpha = (4\pi/\lambda)\,\mathrm{Im}[(\varepsilon_c - \sin^2\theta)^{1/2}]$). For a small oxide layer thickness ($\beta x \ll 1$) and angles θ not too close to 0, or to $\pi/2$, an even simpler approximation for $A_\|(x, \theta)$, similar to equation (6.7) results

$$A_\|(x,\theta) = \frac{A_0}{\cos\theta} + \frac{16\pi k}{n^2\lambda}\frac{\sin^2\theta}{\cos\theta}x. \tag{8.18}$$

Let us confine ourselves to the oxidation activation stage only, and let us assume that only the convective thermal losses are significant (a situation characteristic for the laser heating of metal foils). Then, using equations (6.2), (6.26) and (8.18), one can obtain an adequate expression for evaluating the oxidation activation time

$$t_a^\|(\theta) = fA_0^4 \frac{1}{\cos^2\theta \sin^4\theta} \exp\left(\frac{A_1}{A_0}\sin\theta\right) \tag{8.19}$$

where

$$f = \frac{4\pi}{d_0}\left(\frac{n^2\lambda}{16\pi k A_1}\right)^2 \qquad A_1 = \frac{\eta S T_d}{P}. \tag{8.20}$$

Analysing the expression (8.19), we notice that the activation time $t_a^\|(\theta)$ has a minimum value at θ_{min}, whose value is the solution of the equation

$$\xi\frac{1-\xi^2}{1-3\xi^2} = \frac{2PA_0}{\eta S T_d} \qquad \xi = \cos\theta. \tag{8.21}$$

The value θ_{min} resulting from (8.21) depends on P, and with the rise in the radiation power θ_{min} shifts from values close to 90° to $\theta_{min} = \arccos(1/\sqrt{3}) \simeq 55°$. The maximum gain that could be obtained by the choice of an optimum value of the incidence angle, θ_{min}, can be estimated by means of the expression

$$\left.\frac{1}{k_a^{\parallel}(\theta)}\right|_{\theta=\theta_{min}} = \frac{4\sqrt{2}\pi^2 k^2}{n^4(n^2-1)A_0^2}\xi(1-\xi^2)(1-3\xi^2)\exp\left(\frac{2(1-3\xi^2)}{\xi(1-\xi^2)}\right). \tag{8.22}$$

Several important conclusions can be drawn from equation (8.22), among which we emphasise: (i) the fact that the energy gain obtained as an effect of obliquely incident laser radiation polarised in the incidence plane ($\parallel$) strongly depends on the oxide optical constants and on the initial absorptivity, A_0; and (ii) the fact that this gain can reach, for an optimal choice of the angle of incidence, values as high as $\sim$10–100.

Thus, the theoretical analysis carried out in reference [297] is in good agreement with the experimental results [231], although in the experimental investigations the targets have not been thermally insulated (an approximation implicit in the relation (8.21)). For example the explanation for the lower values of the quantity $1/k_a^{\parallel}(\theta)$ in the case of titanium targets exhibiting high values of the initial absorptivity A_0, by comparison with tungsten samples (situations in which $1/k_a^{\parallel}(\theta) \simeq 1/k_{ig}^{\parallel}(\theta)$), becomes obvious.

Calculations have also indicated [297] that some energy saving can also be obtained at an oblique incidence of the non-polarised radiation, as the rise in $1/k_a^{\parallel}(\theta)$ compensates for the decrease in $1/k_a^{\perp}(\theta)$.

8.2.4. *The radiation wavelength*

It is well known that metal absorptivity increases with decrease of the radiation wavelength, λ, in particular by passing from the mid-IR range, where the radiation of 10.6 μm is situated, to the visible range. Such a behaviour of $A_M(\lambda)$ leads to a shorter time being necessary for heating of the metal up to a certain temperature in a chemically inert atmosphere. Yet, in an oxidising medium, the heating dynamics of metal targets is determined not only by the value of A_M, but also by the optical constants of the oxide. Indeed, the values n and k may exhibit a very intricate dependence on λ (figure 8.7).

As a consequence, one can speak only about a trend of times t_a and t_m to decrease, characteristic for oxidising metal heating, as an effect of using shorter-wavelength lasers, a tendency confirmed by the data reported in reference [242], where $\lambda \simeq 10.6$ μm, 1.06 μm and 0.515 μm cw lasers were used. The same conclusion can also be drawn when analysing many other studies dealing with laser thermal processing of metals [2, 3, 105].

Nevertheless, in certain particular situations, when the laser radiation wavelength is situated in the vicinity of the oxide's absorption bands, certain

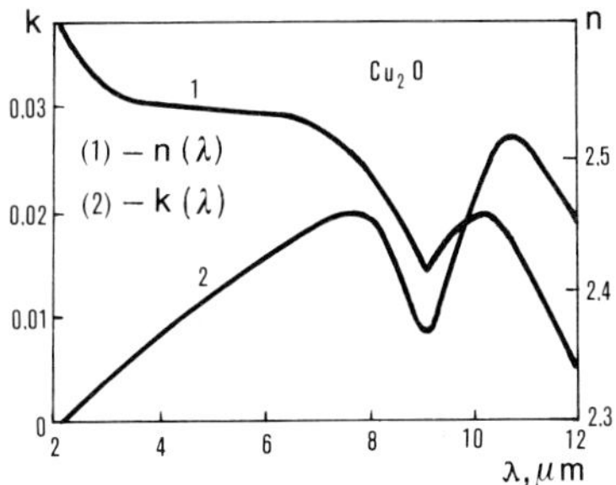

Figure 8.7 Spectral evolutions of the optical constants of the cupric protoxide Cu_2O; (1) $n(\lambda)$, (2) $k(\lambda)$.

anomalous evolutions are possible. For instance, the possibility of some significant minimum values in the evolutions $t_m(\lambda)$ and $E_m(\lambda)$ for the system $Cu + Cu_2O$ has been conjectured in [235]. The spectral dependence of Cu_2O optical constants is given in figure 8.7.

The numerical results, in the case of the function $E_m(\lambda)$, for a copper foil with $m = 50$ mg, $S_s = 0.3$ cm^2, at $P = 20$ W, can be followed in figure 8.8. It is obvious that, featuring a generally decreasing evolution, quantity $E_m(\lambda)$ exhibits a maximum, followed by a minimum at $\lambda \simeq 9.5$ μm. The positions and values of these extreme points depend on the power (or intensity) of the radiation.

The most important practical conclusions that may be drawn from studying the wavelength (frequency) dependence of the efficiency of the metal laser oxidation process can be summarised as follows.

(i) Even for visible laser radiation when the initial absorptivity of metals is relatively high ($A_0 \geqslant 0.1$), the oxidation activation results in an increased rate of sample heating.

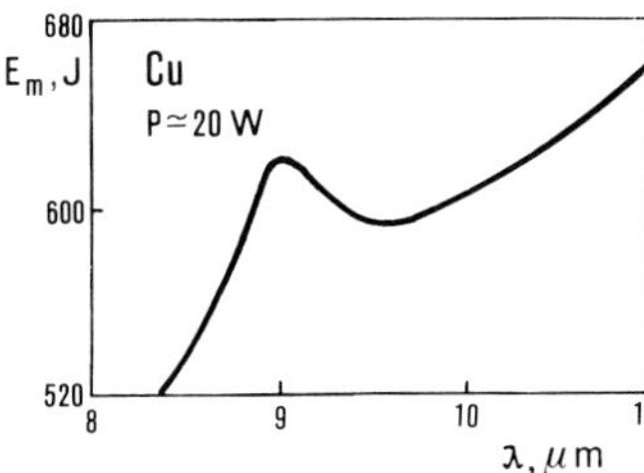

Figure 8.8 Dependence of energy consumption in the heating of a copper target up to the melting point, on the wavelength of the incident radiation; $m = 50$ mg, $P = 20$ W, $S_s = 0.3$ cm^2.

(ii) As CO_2 lasers are intense sources of radiation in the 9–11 μm range, a significant increase in the efficiency of the thermochemical heating regime of the metal target can be obtained and sustained through a careful choice of radiation wavelength, according to the respective values of the optical constants (n and k) of the oxide. Note that, by selecting the frequency, it is possible to obtain with the same CO_2 laser source, the same dissipated power at $\lambda = 9.6$ μm and $\lambda = 10.6$ μm.

References

[1] Arata Y 1985 Development of ultra-high energy density heat source and its application to heat processing *Okada Memorial Lecture* (Japan Society for the Promotion of Welding)

[2] Ready J F 1971 *Effects of High Power Laser Radiation* (New York: Academic)

[3] Duley W W 1976 *CO_2-Lasers: Effects and Applications* (New York: Academic); 1982 *Laser Processing and Analysis of Materials* (New York: Plenum)

[4] Zavecz T E, Saifi M A and Notis M 1975 *Appl. Phys. Lett.* **26** 165

[5] Bonch-Bruevich A M, Imas Ya A, Romanov G S, Libenson M N and Mal'tsev I M 1968 *Sov. Phys.–Tech. Phys.* **38** 851

[6] Basov N G, Boyko A, Krokhin O N, Semenov O G and Sklizkov G V 1968 *Sov. Phys.–Tech. Phys.* **38** 1973

[7] Walters C T 1974 *Appl. Phys. Lett.* **25** 696

[8] Ready J F 1976 *IEEE J. Quantum Electron.* **QE-12** 137

[9] Koo J C and Slusher R E 1976 *Appl. Phys. Lett.* **28** 614

[10] Drude P 1922 *Theory of Optics* (New York: Longmans and Green) (reprint 1966, New York: Dover)

[11] Wooten F 1967 *Optical Properties of Solids* (New York: Academic)

[12] Donovan B 1967 *Elementary Theory of Metals* (New York: Pergamon)

[13] Reuter G E H and Sondheimer E H 1984 *Proc. R. Soc.* A **195** 336

[14] Bennett H E, Silver M and Ashley E J 1963 *J. Opt. Soc. Am.* **53** 1089

[15] Konov V I and Tokarev V N 1983 *Kvant. Elektron.* **10** 327

[16] Schulz I G 1957 *Adv. Phys.* **6** 102

[17] Pippard A B 1947 *Proc. R. Soc.* A **191** 385; 1957 *Phil. Trans. R. Soc.* A **250** 325

[18] Dingle R B 1953 *Physica* **19** 311

[19] Kaganov M I and Azbel' M Yu 1955 *Sov. Phys.–JETP* **29** 49

[20] Azbel' M Yu and Kaner E A 1955 *Sov. Phys.–JETP* **29** 876

[21] Mattis D C and Bardeen J 1958 *Phys. Rev.* **111** 412

[22] Libenson M N and Pudkov S D 1977 *Sov. Phys.–Tech. Phys.* **47** 2441

[23] Sparks M and Loh E Jr 1979 *J. Opt. Soc. Am.* **69** 847

[24] Bennett H E, Bennett J M, Ashley E J and Motyka R H 1968 *Phys. Rev.* **165** 755

[25] Bennett H E and Bennett J M 1966 in *Optical Properties and Electronic Structure of Metals and Alloys* ed F Abeles (Amsterdam: North-Holland) pp 175–188

[26] Abeles F 1972 in *Optical Properties of Solids* ed F Abeles (Amsterdam: North-Holland) p. 93

[27] Theye M L 1970 *Phys. Rev.* B **2** 3060

[28] Nagel S R and Schnatterly S E 1974 *Phys. Rev.* B **9** 1299

[29] Motulevich G P 1971 *Trudy FIAN* (*Moscow*) **55** 3

[30] Pudkov S D 1977 *Sov. Phys.–Tech. Phys.* **47** 649

[31] Bass M and Lion L 1984 *J. Appl. Phys.* **56** 184

[32] Born M and Wolf E 1981 *Principles of Optics* 5th edn (New York: Pergamon)

[33] Haas G and Hadley L 1972 in *American Institute of Physics Handbook* ed D E Gray (New York: McGraw-Hill) p. 6

[34] Weaver J H, Krafka C, Lynch D W and Koch E E 1981 in *Physics Data, Optical Properties of Metals* (Karlsruhe: Fachinformationszentrum)

[35] Ordal M A, Long L L, Bell R J, Bell S E, Bell R R, Alexander L W Jr and Ward C A 1983 *Appl. Opt.* **22** 1099

[36] Bennett J M and Ashley E J 1965 *Appl. Opt.* **4** 221

[37] Porteus J O, Soileau M J and Fontain C W 1976 *Appl. Phys. Lett.* **26** 156
Porteus J O, Soileau M J, Bennett H E and Bass M 1975 *NBS Special Publication* no 435 (Washington, DC: NBS) p. 207

[38] Seitel S C, Porteus J O, Decker D L, Faith W N and Grandjean D J 1981 *IEEE J. Quantum Electron.* **QE-17** 1981 2072

[39] Decker D L and Hodgkin V A 1981 in *Proc. Symp. on Laser Induced Damage in Optical Materials, Boulder, Colorado* ed. H E Bennett *et al*, *NBS Special Publication* no 638 p. 298

[40] Haas G 1965 in *Applied Optics and Optical Engineering* vol. III ed R Kingslake (New York: Academic) p. 309

[41] Arnold G S 1984 *Appl. Opt.* **23** 1435

[42] Elson J M and Bennett J M 1979 *Opt. Eng.* **18** 116

[43] Elson J M and Ritchie R H 1974 *Phys. Status Solidi* **62** 461

[44] Elson J M 1975 *Phys. Rev.* B **12** 2451

[45] Maradudin A and Mills D L 1975 *Phys. Rev.* B **11** 1392

[46] Celli V, Marvin A and Toigo F 1975 *Phys. Rev.* B **11** 1779

[47] Kroger E and Kretschmann E 1976 *Phys. Status Solidi* **76** 515

[48] Sari S O, Kohen D K and Scherkoske K D 1980 *Phys. Rev.* B **21** 2162

[49] Leader J Carl 1979 *J. Opt. Soc. Am.* **69** 610

[50] Elson M, Rahn J P and Bennett J M 1980 *Appl. Opt.* **19** 669

[51] Elson J M and Sung C C 1982 *Appl. Opt.* **21** 1496

[52] Apostol I, Arsenovici L C, Mihailescu I N, Popescu I M, Teodorescu I A and Teodorescu V S 1975 *Rev. Roum. Phys.* **20** 749

[53] Ursu I, Apostol I, Craciun D, Dinescu M, Mihailescu I N, Nistor L C, Popa Al, Teodorescu V S, Prokhorov A M, Chapliev N I and Konov V I 1984 *J. Phys. D: Appl. Phys.* **17** 709

[54] Ursu I, Mihailescu I N, Prokhorov A M and Konov V I 1985 in *Proc. 6th EPS Gen. Conf. 'Trends in Physics', Prague, 1984* vol. 2 ed J Janta and J Pantoflicek p. 442

[55] Walters C T, Barnes R H and Beverly R E III 1978 *J. Appl. Phys.* **49** 2937
Fradin D W and Bass M 1973 *Appl. Phys. Lett.* **22** 157

[56] Danileiko Yu K, Manenkov A A, Prokhorov A M and Khaimov-Mal'kov V Ya 1970 *Sov. Phys.–Tech. Phys.* **58** 31

[57] Bloembergen N 1973 *Appl. Opt.* **12** 661

[58] Ujihara K 1972 *J. Appl. Phys.* **13** 2373

[59] Ursu I, Apostol I, Barbulescu D, Lupei V, Mihailescu I N, Popa Al, Prokhorov A M, Chapliev N I and Konov V I 1983 in *Industrial Applications of Laser Technology* (*Proc. SPIE*, vol. 389) p. 361

Ursu I, Nistor L C, Teodorescu V S, Mihailescu I N, Apostol I, Nanu L, Prokhorov A M, Chapliev N I, Konov V I, Tokarev V N and Ralchenko V G 1983 *Ibid.* p. 398

[60] Namba Y and Tsuwa H 1980 *NBS Special Publication* no 620 (Washington, DC: NBS)

[61] Apollonov V V, Bykovskii Yu A, Degtyarenko N N, Eleshin V Sh, Kozyrev Yu P and Sil'nov S M 1970 *JETP Lett.* **11** 252

[62] Ehler W and Linlor I 1973 *J. Appl. Phys.* **44** 4229

[63] Dinger R, Rohr K and Weber H 1984 *J. Phys. D: Appl. Phys.* **17** 1707

[64] Dinger R, Rohr K and Weber H 1980 *J. Phys. D: Appl. Phys.* **13** 2301

[65] Allen S D, Porteus J O and Faith W N 1982 *Appl. Phys. Lett.* **41** 416

[66] Allen S D, Porteus J O, Faith W N and Franck J B 1984 *Appl. Phys. Lett.* **45** 997

[67] Williams M D 1982 *Appl. Opt.* **21** 747

[68] Berning P H, Hass G and Madden R P 1960 *J. Opt. Soc. Am.* **50** 586

[69] Bolotin G A 1965 *Opt. Spektrosk.* **18** 746

[70] McKay J A and Schriempf J T 1979 *Appl. Phys. Lett.* **35** 433

[71] Koumvakalis N, Lee C S and Bass M 1983 *IEEE J. Quantum Electron.* **QE-19** 1482

[72] Quimby R S, Bass M and Lion L 1981 *Proc. Symp. on Laser Induced Damage in Optical Materials, Boulder, Colorado, 1981* ed. H E Bennett *et al*, *NBS Special Publication* no 638 p. 142

[73] Ursu I, Apostol I, Barbulescu D, Dinescu M, Draganescu V, Mihailescu I N, Moldovan M, Tatu V S, Prokhorov A M, Ageev V P, Konov V I and Tokarev V N 1982 *Rev. Roum. Phys.* **27** 54; 1982 *Appl. Phys.* B **29** 187

[74] Ursu I, Mihailescu I N, Apostol I, Dinescu M, Hening Al, Stoica M, Prokhorov A M, Ageev V P, Konov V I and Tokarev V N 1984 *J. Phys. D: Appl. Phys.* **17** 1315

[75] Ursu I, Mihailescu I N, Prokhorov A M and Konov V I 1984 in *Proc. Int. Conf. 'Lasers and Applications', Bucharest, 1982* vol. 2 (Bucharest: Central Institute of Physics) p. 1

[76] Sanders B A and Gregson V G 1973 in *Proc. Electro-Optical Systems Design Conf., NY, 1973* p.24

[77] Roessler D M and Gregson V G Jr 1978 *Appl. Opt.* **17** 992

[78] Walters C T and Clauer A H 1978 *Appl. Phys. Lett.* **33** 713

[79] McMordie J A and Roberts P D 1975 *J. Phys. D: Appl. Phys.* **8** 768

[80] Porteus J O, Decker D L, Jerningham J L, Faith W N and Bass M 1978 *IEEE J. Quantum Electron.* **QE-14** 776

[81] McKay J A and Schriempf J T 1979 *Appl. Phys. Lett.* **35** 433

[82] Carlsaw H S and Jaeger J C 1959 *Conduction of Heat in Solids* (Oxford: Clarendon)

[83] Bass M (ed) 1983 *Laser Materials Processing* (Amsterdam: North-Holland)

[84] Leybfried E 1963 *Microscopical Theory of the Mechanical and Heat Properties of Materials* (*Glavn. Izd. Fiz. i Mat.*) (Moscow: GIPML)

Stolovich N N and Minitskaya N S 1975 *The Temperature Dependance of the Thermophysical Properties of Several Metals* (Minsk: Nauka)

[85] Ageev V P, Burdin S G, Konov V I, Uglov S A and Chapliev N I 1983 *Kvantov. Elektron.* **10** 780

[86] Ursu I, Apostol I, Barbulescu D, Mihailescu I N, Moldovan M, Prokhorov A M, Ageev V P, Gorbunov A A and Konov V I 1981 *Opt. Commun.* **39** 180; 1981 *Proc. XVth ICPIG, Minsk, 1981* pt II p. 1314

[87] Uglov A A and Isaeva O I 1976 *Fiz. Himia Obrab. Mater.* no 2 23

[88] Rykalin N N, Uglov A A and Nizametdinov M M 1977 *Kvantov. Elektron.* **4** 1509

[89] Vargaftik N V (ed) 1956 *Thermophysical Properties of Substances* (Moscow: Gosenergoizdat)

[90] Lebedev S V, Savvatinskii A N and Sheyndlin M A 1976 *Teplofiz. Vys. Temp.* **14** 285

[91] Libenson M N, Romanov G S and Imas Ya A 1968 *Sov. Phys.–Tech. Phys.* **38** 116

[92] Sparks M and Loh E Jr 1979 *J. Opt. Soc. Am.* **69** 859

[93] Rykalin N N, Uglov A A and Smurov I Yu 1982 *Dokl. Akad. Nauk SSSR* **267** 377

[94] Porteus J O, Decker D L, Faith W N, Grandjean D J, Seitel S C and Soileau M J 1981 *IEEE J. Quant. Electron.* **QE-17** 2078

[95] Rosenthal D 1946 *Trans. Am. Soc. Mech. Eng.* **68** 849

[96] Anisimov S I, Imas Ya A, Romanov G S and Hodyko Yu V 1970 *The Action of High Power Radiation on Metals* (Moscow: Nauka)

[97] Birjega M I, Nanu L, Mihailescu I N, Dinescu M, Popescu-Pogrion N and Sarbu C 1986 *Optica Acta* **33** 1073; 1986 *Phys. Status Solidi* a **35** 423

[98] Brugger K 1972 *J. Appl. Phys.* **43** 577

[99] Gonsalves J N and Duley W W 1972 *Can. J. Phys.* **43** 4684

[100] Lin T P 1967 *IBM J. Res. Dev.* **11** 527

[101] Sparks M 1976 *J. Appl. Phys.* **47** 837

[102] Stern F 1973 *J. Appl. Phys.* **44** 4204

[103] Apostol I, Cojocaru E, Draganescu V, Mihailescu I N, Nistor L C and Teodorescu V S 1981 *Rev. Roum. Phys.* **26** 357

[104] Ursu I, Apostol I, Cojocaru E, Mihailescu I N, Nistor L C and Teodorescu V S 1981 *Proc. 5th EPS Gen. Conf. 'Trends in Physics', Istanbul, 1981* (Geneva: EPS) p. 267

[105] Apollonov V V, Barchukov A I, Konyukhov V K and Prokhorov A M 1972 *Sov. Phys.–JETP* **15** 248

[106] Apollonov V V, Barchukov A I and Prokhorov A M 1974 *IEEE J. Quantum Electron.* **QE-10** 505

[107] Apollonov V V, Prokhorov A M, Khomich V Yu and Chetkin S A 1983 *NBS Special Publication* no 735 (Washington, DC: NBS)

[108] Apollonov V V, Prokhorov A M, Khomich V Yu and Chetkin S A 1982 *Kvantov. Elektron.* **9** 343

[109] Novatskii V 1962 *Problems of Thermoelasticity* (Moscow: Izd. AN SSSR)

[110] Parkus G 1963 *Permanent Temperature Stresses* (Moscow: Fizmatghiz)

[111] Apollonov V V, Barchukov A I, Karlov N V, Prokhorov A M, Khomich V Yu and Shefter E M 1975 *Sov. Phys.–Tech. Phys.* **1** 522

[112] Apollonov V V, Prokhorov A M, Khomich V Yu and Chetkin S A 1981 *Kvantov. Elektron.* **8** 2208
[113] Apollonov V V, Bunkin F V, Khomich V Yu and Chetkin S A 1978 *Sov. Phys.–Tech. Phys.* **4** 1017
[114] Ekordi T 1971 *Physics and Mechanics of Damage and Stability of Solids* (Moscow: Nauka)
[115] Lee C S, Koumvakalis N and Bass M 1983 *J. Appl. Phys.* **54** 5727
[116] Jacobson D L, Bickford W, Kidd J, Barthelmy R and Bloomer R H 1975 *AIAA Paper* no 719
[117] Apollonov V V, Barchukov A I, Ostrovskaya L M, Rodin V N, Khomich V Yu and Tsypin M I 1978 *Kvantov. Elektron.* **5** 446
[118] Apollonov V V, Barchukov A I, Borodin V I, Bystrov V I, Goncharov V N, Ostrovskaya L M, Prokhorov A M, Rodin V N, Trushin E V, Khomich V Yu, Tsypin M I, Shevyakhin Yu F and Shur Ya Sh 1978 *Kvantov. Elektron.* **5** 1169
[119] Porteus J O, Decker D L, Grandjean D J, Seitel S C and Faith W 1980 *NBS Special Publication* no 568 (Washington, DC: NBS) p. 175
[120] Konov V I, Pimenov S M, Prokhorov A M and Chapliev N I 1987 *Poverhnost'* **12** 98
[121] Dobrovol'skii I P and Uglov A A 1974 *Kvantov. Elektron.* **4** 788
[122] Thomas S J, Harrison R F and Figueira J F 1982 *Appl. Phys. Lett.* **40** 200
[123] Figueira J F and Thomas S J 1983 in *Surface Studies with Lasers* ed F R Aussenegg, A Leitner and M E Lippitsch (Berlin: Springer) p. 212
[124] Figueira J F and Thomas S J 1982 *IEEE J. Quantum Electron.* **QE-18** 1380
[125] Goldstein I, Bua D and Horrigan F A 1975 *NBS Special Publication* no 435 (Washington, DC: NBS) p. 41
[126] Adams C and Hardway G 1965 *IEEE Trans. Ind. Gen. Appl.* **IGA-1** 90
[127] Anisimov S I, Bonch-Bruevich A M, El'yashevich M A, Imas Ya A, Pavlenko N A and Romanov G S 1967 *Sov. Phys.–Tech. Phys.* **11** 945
[128] Ready J F 1974 *Appl. Phys. Lett.* **25** 558
[129] Ursu I, Apostol I, Dinescu M, Mihailescu I N, Popa Al, Prokhorov A M, Konov V I and Chapliev N I 1984 *Appl. Phys.* A **34** 133
[130] Lee C S, Koumvakalis N and Bass M 1982 *Appl. Phys. Lett.* **47** 625
[131] Nielsen P D 1975 *J. Appl. Phys.* **46** 4501
[132] Arkhipov Yu V, Belashkov I N, Datskevich N P, Egorov V N, Izyumov A F, Karlov N V, Konov V I, Kononov N N, Kuz'min GP, Nesterenko A A and Chapliev N I 1986 *Kvantov. Elektron.* **13** 103
[133] Bastow T J 1969 *Nature* **222** 1058
[134] Ursu I, Mihailescu I N, Popa Al, Prokhorov A M, Ageev V P, Gorbunov A A and Konov V I 1985 *J. Appl. Phys.* **58** 3909
[135] Siegrist M, Kaech G and Kneubuhl F K 1973 *Appl. Phys.* **2** 45
[136] Birnbaum M J 1965 *J. Appl. Phys.* **36** 3688
[137] Emmony D C, Howson R P and Willis L J 1973 *Appl. Phys. Lett.* **23** 598
[138] Willis L J and Emmony D C 1975 *Opt. Laser Technol.* (Oct.) 222
[139] Isenor N R 1977 *Appl. Phys. Lett.* **31** 148
[140] Leamy H J, Rozganyi J A, Sheng T T and Celler J K 1978 *Appl. Phys. Lett.* **32** 535
[141] Oron M and Sorensen J 1979 *Appl. Phys. Lett.* **35** 782

[142] Jain A K, Kulkarni V N, Sood D K and Uppal J S 1981 *J. Appl. Phys.* **52** 4882
[143] Fauchet P M and Siegmann A E 1983 *Appl. Phys.* A **32** 135
[144] Temple P A and Soileau M J 1981 *IEEE J. Quantum Electron.* **QE-17** 2067
[145] Anisimov V N, Baranov V Yu, Bol'shov L A, Dyhne A M, Malyuta D D, Pis'mennyi V D, Sebrant A Yu and Stepanova M A 1983 *Poverhnost'* **7** 138
[146] Soileau M J and Stryland E W 1982 *NBS Report* no 669 406
[147] van Vechten J A 1981 *Solid State Commun.* **39** 1285
[148] Keilmann F and Bai Y H 1982 *Appl. Phys.* A **29** 9
[149] Aksenov V P and Jurkyn V G 1982 *Dokl. Akad. Nauk SSSR* **265** 1365
[150] Young J F, Sipe J E, Preston J S and van Driel H M 1982 *Appl. Phys. Lett.* **41** 261
[151] Young J F, Preston J S, van Driel H M and Sipe J E 1983 *Phys. Rev.* B **27** 1155
[152] Prokhorov A M, Svakhin A S, Sychugov V A, Tischenko A V and Hakimov A A 1983 *Kvantov. Elektron.* **10** 906
[153] Young J F, Sipe J E and van Driel H M 1984 *Phys. Rev.* B **30** 2001
[154] van Driel H M, Sipe J E and Young J F 1982 *Phys. Rev. Lett.* **49** 1955
[155] Aksenov V P 1982 *FIAN Preprint* no 194 (Moscow: FIAN)
[156] Bonch-Bruevich A M, Kochenghina M K, Libenson M N and Makyn V S 1982 *Izv. Akad. Nauk SSSR, Ser. Fiz.* **46** 1186
[157] Ehrlich D J, Brueck S R and Tsao J Y 1982 *Appl. Phys. Lett.* **41** 630
[158] Keilmann F 1983 *Phys. Rev. Lett.* **51** 2097
[159] Konov V I, Prokhorov A M, Sychugov V A and Tokarev V N 1985 *Poverhnost'* no 1 128
[160] Sipe J E, Young J F, Preston J S and van Driel H M 1983 *Phys. Rev.* B **27** 1141
[161] Sychugov V A and Tulaykova T V 1984 *Kvantov. Elektron.* **11** 437
[162] Guosheng Zhou, Fouchet P M and Siegmann A E 1982 *Phys. Rev.* B **21** 5366
[163] Avrutskii I A, Bazakutsa P V, Prokhorov A M and Sychugov V A 1985 *Kvantov. Elektron.* **12** 650
[164] Prokhorov A M, Sychugov V A, Tischenko A V and Hakimov A A 1982 *Sov. Phys.–Tech. Phys.* **8** 1409
[165] Bonch-Bruevich A M, Libenson M N and Makyn V S 1984 *Sov. Phys.–Tech. Phys.* **10** 3
[166] Bazakutsa P V, Maslennikov V L, Prokhorov A M, Sychugov V A and Tischenko A V 1984 *Kvantov. Elektron.* **11** 1447
[167] Bazakutsa P V, Sychugov V A and Prokhorov A M 1984 *Kvantov. Elektron.* **11** 2127
[168] Emel'yanov V I, Zemskov E M and Seminogov V N 1984 *Kvantov. Elektron.* **11** 2283
[169] Emel'yanov V I, Konov V I, Seminogov V N and Tokarev V N 1987 *Preprint* no 89 (Moscow: IOFAN SSSR)
[170] Avrutskii I A, Bazakutsa P V, Prokhorov A M, Sychugov V A and Tischenko A V 1984 *Sov. Phys.–Tech. Phys.* **10** 1086
[171] Maracas J N, Harris J L, Lee C A and McFarlane R A 1978 *Appl. Phys. Lett.* **33** 453
[172] Hill C and Godfray D J 1980 *J. Physique* **41** suppl. 5 C4-79
[173] Emel'yanov V I, Seminogov V N and Zemskov E M 1983 *Kvantov. Elektron.* **10** 2389
[174] Emel'yanov V I and Seminogov V N 1984 *Kvantov. Elektron.* **11** 871

[175] Akhmanov S A, Emel'yanov V I, Koroteev N I and Seminogov V N 1985 *Usp. Fiz. Nauk* **147** 675
[176] Emel'yanov V I and Seminogov V N 1984 *Sov. Phys.–JETP* **86** 1026
[177] Samokhin A A 1983 *Kvantov. Elektron.* **10** 2022
[178] Tribel'skii M I and Gol'berg S M 1982 *Sov. Phys.–Tech. Phys.* **8** 1227
[179] Ursu I, Mihailescu I N, Nistor L C, Teodorescu V S, Prokhorov A M, Konov V I and Tokarev V N 1985 *Appl. Opt.* **24** 3336
[180] Ursu I, Mihailescu I N, Prokhorov A M, Konov V I and Tokarev V N 1985 *Physica* **132C** 395
[181] Ursu I, Mihailescu I N, Prokhorov A M, Konov V I and Tokarev V N *J. Appl. Phys.* **61** 2445
[182] Ursu I, Mihailescu I N, Prokhorov A M, Konov V I and Tokarev V N *Europhys. Lett.* **2** 685
[183] Ursu I, Mihailescu I N, Popa Al, Prokhorov A M, Konov V I, Ageev V P and Tokarev V N 1984 *Appl. Phys. Lett.* **45** 365
[184] Nevière M and Reinish R 1982 *Phys. Rev.* B **26** 5403
[185] Garcia N, Diaz G, Saenz J J and Ocal C 1984 *Surf. Sci.* **143** 342
[186] Ermakov A A, Konov V I, Nikitin P I, Prokhorov A M, Uglov S A and Shabanov A R 1986 *Preprint* no 357 (Moscow: IOFAN)
[187] Sullivan A and Houldcroft P 1967 *Br. Weld. J.* no 7 443
[188] Barchukov A I and Mirkin L I 1967 *Fiz. Himia Obrab. Mater.* no 6 126
[189] Asmus F T and Baker F S 1969 in *Records of the 11th Symp. on Electron, Ion and Laser Beam Technology* (IEEE: Michigan) p. 241
[190] Gonsalves J N and Duley W W 1971 *Can. J. Phys.* **49** 1708
[191] Veiko V P and Libenson M N 1977 *Laser Processing* (Leningrad: Lenizdat)
[192] Volod'kina V I, Krylov K I and Libenson M N 1973 *Fiz. Himia Obrab. Mater.* no 5 145
[193] Veiko V P, Kotov G A, Libenson M N and Nikitin M N 1973 *Dokl. Akad. Nauk SSSR* **208** 587
[194] Rykalin N N, Uglov A A and Kokora A N 1974 *Fiz. Himia Obrab. Mater.* no 1 3
[195] Barchukov A I, Bunkin F V, Konov V I and Prokhorov A M 1973 *FIAN Preprint* no 165 (Moscow: FIAN)
[196] Arzuov M I, Barchukov A I, Bunkin F V, Konov V I and Lyubin A A 1975 *Kvantov. Elektron.* **2** 1717
[197] Hauffe K 1963 *Reactions in Solids and on Their Surface* (Moscow: Innostranaya Literatura)
[198] Kubashevskii O and Hopkins B 1965 *Oxidation of Metals and Alloys* (Moscow: Metallurghia)
[199] Koffstadt P 1969 *High Temperature Oxidation of Metals* (Moscow: Mir)
[200] Koffstadt P 1975 *Deviation from Stoichiometry, Diffusion and Electroconduction in Simple Metal Oxides* (Moscow: Mir)
[201] Bénard J 1968 *Oxidation of Metals* (Moscow: Metallurghia)
[202] Bunkin F V, Kirichenko N A and Luk'yanchuk B S 1982 *Usp. Fiz. Nauk* **138** 45
[203] Bonch-Bruevich A M and Libenson M N 1982 *Izv. Akad. Nauk SSSR, Ser. Fiz.* **46** 1104
[204] Konov V I 1983 in *Proc. Int. Conf. and School on Lasers and Appl. LAICS '82, Bucharest* pt 1 p. 665

[205] Luk'yanchuk B S 1983 *Proc. Int. Conf. and School on Lasers and Appl. LAICS '82, Bucharest* pt 1 p. 771
[206] Burmistrov A V 1981 *Sov. Phys.–Tech. Phys.* **25** 1733
[207] Bunkin F V, Kirichenko N A and Luk'yanchuk B S 1981 *Izv. Akad. Nauk SSSR, Ser. Fiz.* **45** 1018
[208] Arzuov M I, Barchukov A I, Bunkin F V, Kirichenko N A, Konov V I and Luk'yanchuk B S 1979 *Kvantov. Elektron.* **6** 466
[209] Duley W W, Semple D, Morency J and Gravel M 1979 *Opt. Laser Technol.* (Nov.) 313
[210] Buzykin O G, Burmistrov A V, Klynkin S S, Kogan M N and Ushkov V M 1982 *Dokl. Akad. Nauk* **263** 1115
[211] Arzuov M I, Konov V I and Metev S M 1978 *Fiz. Himia Obrab. Mater.* no 5 19
[212] Arzuov M I, Karasev M E, Konov V I, Kostin V V, Metev S M, Silenok A S and Chapliev N I 1978 *Kvantov. Elektron.* **5** 1567
[213] Buzykin O G and Burmistrov V A 1982 *Poverhnosti* no 2 63
[214] Konov V I, Ralchenko V G and Tokarev V N 1982 *Izv. Akad. Nauk SSSR, Ser. Fiz.* **46** 1065
[215] Dgugunovich V A, Zhdanovski V A and Snopko V N 1981 *J. Prikl. Spektrosk.* **34** 799
[216] Ursu I, Nistor L C, Teodorescu V S, Mihailescu I N, Nanu L, Prokhorov A M, Chapliev N I and Konov V I 1984 *Appl. Phys. Lett.* **44** 188
[217] Pasternak J 1961 *Czech. J. Phys.* B **11** 374
[218] Oron M, Sendsen L G and Sorensen G 1979 *NBS Special Publication* no 568 (Washington, DC: NBS)
[219] Konov V I, Prokhorov A M, Ralchenko V G, Stepanov Yu I, Shirkov N I and Shtamchaev M I 1985 *Kratk. Soobsch. Fiz.* no 9 3
[220] Nefedov V I and Cherepin V G 1983 *Physical Methods for the Investigation of Solid Surfaces* (Moscow: Nauka)
[221] Metev S M, Savtchenko S I, Stamenov K V, Veiko V P and Kotov G A 1981 *IEEE J. Quantum Electron.* **QE-17** 2004
[222] Metev S M, Veiko V P, Stamenov K V and Kalev Kh A 1977 *Kvantov. Elektron.* **7** 863
[223] Ursu I, Nanu L, Mihailescu I N, Nistor L C, Teodorescu V S, Prokhorov A M, Konov V I and Chapliev N I 1984 *J. Physique Lett.* **45** 737
Ursu I, Mihailescu I N and Nanu L 1987 *Appl. Phys. Lett.* **49** 109
[224] Burmistrov A V and Konov V I 1982 *Fiz. Himia Obrab. Mater.* no 3 3
[225] Campbell W E and Thomas U V 1947 *Trans. Electrochem. Soc.* **91** 628
[226] Arzuov M I, Barchukov A I, Bunkin F V, Konov V I and Luk'yanchuk B S 1979 *Kvantov. Elektron.* **6** 1339
[227] Boyko V I, Kirichenko N A, Konov V I, Luk'yanchuk B S, Nanai L, Simakhin A V, Tokarev V N, Hevesi I and Shafeev G A 1982 *Kratk. Soobsch. Fiz.* no 9 49
[228] Bunkin F V, Kirichenko N A, Luk'yanchuk B S, Simakhin A V, Shafeev G A, Nanai L and Hevesi I 1983 *Acta Phys. Hungar.* **54** 111
[229] Ursu I, Nanu L, Dinescu M, Hening Al, Mihailescu I N, Nistor L C, Teodorescu V S, Szil E, Hevesi I, Kovacs J and Nanai L 1984 *Appl. Phys.* A **35** 103

[230] Bonch-Bruevich A M, Libenson M N, Makin V A, Pudkov S D, Ivanova M N and Kochenghina M K 1978 *Sov. Phys.–Tech. Phys.* **4** 921

[231] Goncharov I N, Gorbunov A A, Konov V I, Silenok A S, Skvortsov Yu A, Tokarev V N and Chapliev N I 1980 *FIAN Preprint* no 76 (Moscow: FIAN)

[232] Boiko V I, Bunkin F V, Kirichenko N A and Luk'yanchuk B S 1980 *Dokl. Akad. Nauk SSSR* **250** 78

[233] Akimov A G, Gagarin A P, Danghurov V G, Makin V A and Pudkov S D 1980 *Sov. Phys.–Tech. Phys.* **50** 2461

[234] Mott N F 1979 *The Transition Metal–Insulator* (Moscow: Nauka)

[235] Akimov A G, Bonch-Bruevich A M, Gagarin A P, Dorofeev V G, Libensen M N, Makin V S and Pudkov S D 1980 *Sov. Phys.–Tech. Phys.* **6** 1017

[236] 1976 *Handbook of Electrotechnical Materials* vol. 3 (Leningrad: Energhia)

[237] Arzuov M I, Konov V I, Kostin V V, Metev S M, Silenok A S and Chapliev N I 1977 *FIAN Preprint* no 152 (Moscow: FIAN)

[238] Arzuov M I, Bunkin F V, Kirichenko N A, Konov V I and Luk'yanchuk B S 1978 *FIAN Preprint* no 39 (Moscow: FIAN)

[239] Libenson M N 1978 *Sov. Phys.–Tech. Phys.* **4** 917

[240] Uglov A A, Smurov Yu I and Volkov A A 1983 *Kvantov. Elektron.* **10** 289

[241] Arzuov M I, Bunkin F V, Kirichenko N A, Konov V I and Luk'yanchuk B S 1978 *Sov. Phys.–Tech. Phys.* **4** 230

[242] Bunkin F V, Kirichenko N A, Konov V I and Luk'yanchuk B S 1980 *Kvantov. Elektron.* **7** 1548

[243] Akimov A G, Bonch-Bruevich A M, Gagarin A P, Dorofeev V G, Zimin N A, Ivanova I N, Libenson M N, Makin V S and Pudkov S D 1982 *Izv. Akad. Nauk SSSR, Ser. Fiz.* **46** 1177

[244] Bunkin F V, Kirichenko N A, Luk'yanchuk B S and Minervina O I 1980 *Sov. Phys.–Tech. Phys.* **6** 101

[245] Bonch-Bruevich A M, Libenson M N, Makin V S and Pudkov S D 1977 *Sov. Phys.–Tech. Phys.* **3** 193

[246] Bonch-Bruevich A M, Dorofeev V G, Libenson M N, Makin V S, Pudkov S D and Rubonova G M 1982 *Sov. Phys.–Tech. Phys.* **52** 1133

[247] Crane K C A, Garnsworthy R K and Matias L E S 1980 *J. Appl. Phys.* **51** 5954

[248] Bobyrev V A, Bunkin F V, Kirichenko N A, Luk'yanchuk B S and Simakhin A V 1980 *JETP Lett.* **32** 608

[249] Bobyrev V A, Bunkin F V, Kirichenko N A, Luk'yanchuk B S and Simakhin A V 1982 *Kvantov. Elektron.* **9** 692

[250] Bobyrev V A, Bunkin F V, Kirichenko N A, Luk'yanchuk B S and Simakhin A V 1983 *Kvantov. Elektron.* **10** 793

[251] Prokhorov A M, Buzykin O G, Burmistrov A V, Kogan M N, Kogan V I, Simakhin A V and Ralchenko V G 1983 *Dokl. Akad. Nauk* **271** 1126

[252] Buzykin O G, Burmistrov A V, Kogan M N, Konov V I, Prokhorov A M and Ralchenko V G 1983 *FIAN Preprint* no 212 (Moscow: FIAN)

[253] Prokhorov A M, Buzykin O G, Burmistrov A V, Kogan M N, Konov V I and Ralchenko V G 1984 *Dokl. Akad. Nauk* **274** 1357

[254] Arzuov M I, Borodatov S A, Buzykin O G, Burmistrov A V, Konov V I,

Mihailescu I N, Popa Al, Popescu M, Prokhorov A M, Ralchenko V G and Chapliev N I 1985 *Poverhnost'* no 5 130

[255] Ursu I, Mihailescu I N, Nanu L, Prokhorov A M, Konov V I and Ralchenko V G 1985 *Appl. Phys. Lett.* **46** 110

[256] Ageev V P, Arzuov M I, Konov V I, Silenok A S and Chapliev N I 1977 *Sov. Phys.–Tech. Phys.* **3** 1179

[257] Savinich V S 1983 *Sov. Phys.–Tech. Phys.* **53** 2253

[258] Bunkin F V, Kirichenko N A and Luk'yanchuk B S 1982 *Kvantov. Elektron.* **9** 1959

[259] Buzykin O G and Burmistrov A V 1982 *Sov. Phys.–Tech. Phys.* **8** 744

[260] Anisimov N R 1982 *Sov. Phys.–Tech. Phys.* **8** 1320

[261] Burmistrov A V 1982 *Sov. Phys.–Tech. Phys.* **8** 29

[262] Samsonov S V (ed) 1978 *Physicochemical Properties of Oxides* (Moscow: Metallurghia)

[263] Bunkin F V, Kirichenko N A, Luk'yanchuk B S and Shafeev G A 1982 *Kvantov. Elektron.* **9** 1848

[264] Spitzer W G, Miller R C, Kleinman D A and Howarth L E 1962 *Phys. Rev.* **126** 1710

[265] Bai A S, Lainer D M, Slyusareva E N and Tsipin M I 1970 *Oxidation of Titanium and its Alloys* (Moscow: Metallurghia)

[266] Kikoin I K (ed) 1976 *Tables of Physical Constants* (Moscow: Atomizdat)

[267] Bobyrev V A, Bunkin F V, Kirichenko N A, Luk'yanchuk B S, Simakhin A V and Shafeev G A 1984 *Poverhnost'* no 4 134

[268] Voytovich A F 1968 *Oxidation of Carbides and Nitrides* (Kiev: Naukova Dumka)

Ursu I, Mihailescu I N, Gutu I, Hening Al, Julea Th, Nistor L C, Teodorescu V S, Prokhorov A M, Konov V I and Ralchenko V G 1986 *Appl. Opt.* **25** 2720

[269] Ursu I, Mihailescu I N, Nistor L C, Teodorescu V S, Prokhorov A M, Konov V I and Ralchenko V G 1987 *Appl. Phys. Lett.* **50** 563

[270] Konov V I, Ralchenko V G and Mihailescu I N 1987 in *Proc. Int. Conf. on Laser Advanced Materials Processing: Science and Applications* (*Osaka, Japan*) ed. Y Arata (Osaka: Welding Institute) p. 477

[271] Bonch-Bruevich A M, Dorofeev V G, Libenson M N, Makin V S and Pudkov S D 1982 *Sov. Phys.–Tech. Phys.* **8** 1217

[272] Boiko V I, Kirichenko N A and Luk'yanchuk B S 1982 *FIAN Preprint* no 31 (Moscow: FIAN)

[273] Alimov D T, Atabaev Sh, Bunkin F V, Juravskii V L, Kirichenko N A, Luk'yanchuk B S, Omel'chenko A I, Simakhin A V and Habibulaev P K 1983 *Dokl. Akad. Nauk* **268** 850

[274] Strakhovskii L G 1982 *Fiz. Gorenyia i Vzryva* no 4 92

[275] Gol'berg S M and Tribel'skii M I 1982 *Sov. Phys.–Tech. Phys.* **8** 178

[276] Gol'berg S M, Matyushin G A and Tribel'skii M I 1983 *Poverhnost'* no 9 136

[277] Buzykin O G, Burmistrov A V, Volod'kina V L, Gol'berg S M, Matyushin G A and Tribel'skii M I 1982 *Poverhnost'* no 10 117

[278] Alimov D T, Atabaev Sh, Bunkin F V, Juravskii V L, Kirichenko N A, Luk'yanchuk B S, Omel'chenko A I and Habibulaev P K 1982 *Poverhnost'* no 8 12

[279] Alimov D T, Bunkin F V, Edravskii I D, Kirichenko N A, Luk'yanchuk B S and Habibulaev P K 1982 *Poverhnost'* no 9 82

[280] Buzykin O G, Burmistrov A K, Klinkin S S, Kogan M N and Yuzov V M 1982 *Dokl. Akad. Nauk SSSR* **263** 1115

[281] Kirichenko N A and Luk'yanchuk B S 1983 *Kvantov. Elektron.* **10** 819

[282] Bunkin F V, Kirichenko N A and Luk'yanchuk B S 1982 *Kvantov. Elektron.* **9** 1959

[283] Buzykin O G, Burmistrov A V and Kogan M N 1982 *Poverhnost'* no 9 91

[284] Alimov D T, Atabaev Sh, Bunkin F V, Juravskii V, Kirichenko N A, Luk'yanchuk B S, Omel'chenko A I and Habibulaev P K 1982 *Sov. Phys.–Tech. Phys.* **8** 10

[285] Bobyrev V A, Bunkin F V, Deli E, Kirichenko N A, Luk'yanchuk B S, Nanai L, Simakhin A V, Hevesi I and Shafeev G A 1982 *Kvantov. Elektron.* **9** 1943

[286] Stel'makh M F (ed) 1975 *Lasers in Technology* (Moscow: Energhia)

[287] Bunkin F V, Kirichenko N A and Luk'yanchuk B S 1978 *FIAN Preprint* no 146 (Moscow: FIAN)

[288] Pankova M B and Savinich V S 1977 *Fiz. Himia Obrab. Mater.* no 1 33

[289] Savinich V S 1977 *Fiz. Himia Obrab. Mater.* no 1 37

[290] Arzuov M I, Bunkin F V, Kirichenko N A, Konov V I and Luk'yanchuk B S 1978 *Kratk. Soobsch. Fiz.* no 11 43

[291] Bunkin F V, Kirichenko N A and Luk'yanchuk B S 1980 *Fiz. Himia Obrab. Mater.* no 5 7

[292] Arzuov M I, Bunkin F V, Kirichenko N A, Konov V I and Luk'yanchuk B S 1978 *Kratk. Soobsch. Fiz.* no 11 43

[293] Bunkin F V, Kirichenko N A, Kraskov I V, Luk'yanchuk B S and Shkedov I M 1983 *Dokl. Akad. Nauk SSSR* **268** 5198

[294] Arzuov M I, Bunkin F V, Kirichenko N A, Konov V I and Luk'yanchuk B S 1979 *Sov. Phys.–Tech. Phys.* **5** 193

[295] Volod'kina V L and Kotov G A 1982 *Sov. Phys.–Tech. Phys.* **52** 64

[296] Ready J F 1970 *Laser Focus* **6** 38

[297] Arzuov M I, Barchukov A I, Bunkin F V, Kirichenko N A, Konov V I and Luk'yanchuk B S 1979 *Kvantov. Elektron.* **6** 1432

Index